Reviews of Environmental Contamination and Toxicology

VOLUME 195

Reviews of Environmental Contamination and Toxicology

Editor
David M. Whitacre

VOLUME 195

Springer

Coordinating Board of Editors

Springer
New York: 233 Spring Street, New York, NY 10013, USA
Heidelberg: Postfach 10 52 80, 69042 Heidelberg, Germany

ISSN 0179-5953

Printed on acid-free paper.

ISBN: 978-1-4419-2645-6 e-ISBN: 978-0-387-77030-7
DOI 10.1007/978-0-387-77030-7

springer.com

Foreword

International concern in scientific, industrial, and governmental communities over traces of xenobiotics in foods and in both abiotic and biotic environments has justified the present triumvirate of specialized publications in this field: comprehensive reviews, rapidly published research papers and progress reports, and archival documentations. These three international publications are integrated and scheduled to provide the coherency essential for nonduplicative and current progress in a field as dynamic and complex as environmental contamination and toxicology. This series is reserved exclusively for the diversified literature on "toxic" chemicals in our food, our feeds, our homes, recreational and working surroundings, our domestic animals, our wildlife and ourselves. Tremendous efforts worldwide have been mobilized to evaluate the nature, presence, magnitude, fate, and toxicology of the chemicals loosed upon the earth. Among the sequelae of this broad new emphasis is an undeniable need for an articulated set of authoritative publications, where one can find the latest important world literature produced by these emerging areas of science together with documentation of pertinent ancillary legislation.

Research directors and legislative or administrative advisers do not have the time to scan the escalating number of technical publications that may contain articles important to current responsibility. Rather, these individuals need the background provided by detailed reviews and the assurance that the latest information is made available to them, all with minimal literature searching. Similarly, the scientist assigned or attracted to a new problem is required to glean all literature pertinent to the task, to publish new developments or important new experimental details quickly, to inform others of findings that might alter their own efforts, and eventually to publish all his/her supporting data and conclusions for archival purposes.

In the fields of environmental contamination and toxicology, the sum of these concerns and responsibilities is decisively addressed by the uniform, encompassing, and timely publication format of the Springer triumvirate:

Reviews of Environmental Contamination and Toxicology [Vol. 1 through 97 (1962–1986) as Residue Reviews] for detailed review articles concerned with any aspects of chemical contaminants, including pesticides, in the total environment with toxicological considerations and consequences.

Bulletin of Environmental Contamination and Toxicology (Vol. 1 in 1966) for rapid publication of short reports of significant advances and discoveries in the fields of air, soil, water, and food contamination and pollution as well as methodology and other disciplines concerned with the introduction, presence, and effects of toxicants in the total environment.

Archives of Environmental Contamination and Toxicology (Vol. 1 in 1973) for important complete articles emphasizing and describing original experimental or theoretical research work pertaining to the scientific aspects of chemical contaminants in the environment.

Manuscripts for *Reviews* and the *Archives* are in identical formats and are peer reviewed by scientists in the field for adequacy and value; manuscripts for the *Bulletin* are also reviewed, but are published by photo-offset from camera-ready copy to provide the latest results with minimum delay. The individual editors of these three publications comprise the joint Coordinating Board of Editors with referral within the Board of manuscripts submitted to one publication but deemed by major emphasis or length more suitable for one of the others.

Coordinating Board of Editors

Preface

The role of *Reviews* is to publish detailed scientific review articles on all aspects of environmental contamination and associated toxicological consequences. Such articles facilitate the often-complex task of accessing and interpreting cogent scientific data within the confines of one or more closely related research fields.

In the nearly 50 years since *Reviews of Environmental Contamination and Toxicology* (formerly *Residue Reviews*) was first published, the number, scope and complexity of environmental pollution incidents have grown unabated. During this entire period, the emphasis has been on publishing articles that address the presence and toxicity of environmental contaminants. New research is published each year on a myriad of environmental pollution issues facing peoples worldwide. This fact, and the routine discovery and reporting of new environmental contamination cases, creates an increasingly important function for *Reviews*.

The staggering volume of scientific literature demands remedy by which data can be synthesized and made available to readers in an abridged form. *Reviews* addresses this need and provides detailed reviews worldwide to key scientists and science or policy administrators, whether employed by government, universities or the private sector.

There is a panoply of environmental issues and concerns on which many scientists have focused their research in past years. The scope of this list is quite broad, encompassing environmental events globally that affect marine and terrestrial ecosystems; biotic and abiotic environments; impacts on plants, humans and wildlife; and pollutants, both chemical and radioactive; as well as the ravages of environmental disease in virtually all environmental media (soil, water, air). New or enhanced safety and environmental concerns have emerged in the last decade to be added to incidents covered by the media, studied by scientists, and addressed by governmental and private institutions. Among these are events so striking that they are creating a paradigm shift. Two in particular are at the center of ever-increasing media as well as scientific attention: bioterrorism and global warming. Unfortunately, these very worrisome issues are now super-imposed on the already extensive list of ongoing environmental challenges.

The ultimate role of publishing scientific research is to enhance understanding of the environment in ways that allow the public to be better informed. The term "informed public" as used by Thomas Jefferson in the age of enlightenment conveyed the thought of soundness and good judgment. In the modern sense, being "well informed" has the narrower meaning of having access to sufficient information. Because the public still gets most of its information on science and technology from TV news and reports, the role for scientists as interpreters and brokers of scientific information to the public will grow rather than diminish.

Environmentalism is the newest global political force, resulting in the emergence of multi-national consortia to control pollution and the evolution of the environmental ethic. Will the new politics of the 21st century involve a consortium of technologists and environmentalists, or a progressive confrontation? These matters are of genuine concern to governmental agencies and legislative bodies around the world.

For those who make the decisions about how our planet is managed, there is an ongoing need for continual surveillance and intelligent controls, to avoid endangering the environment, public health, and wildlife. Ensuring safety-in-use of the many chemicals involved in our highly industrialized culture is a dynamic challenge, for the old, established materials are continually being displaced by newly developed molecules more acceptable to federal and state regulatory agencies, public health officials, and environmentalists.

Reviews publishes synoptic articles designed to treat the presence, fate, and, if possible, the safety of xenobiotics in any segment of the environment. These reviews can either be general or specific, but properly lie in the domains of analytical chemistry and its methodology, biochemistry, human and animal medicine, legislation, pharmacology, physiology, toxicology and regulation. Certain affairs in food technology concerned specifically with pesticide and other food-additive problems may also be appropriate.

Because manuscripts are published in the order in which they are received in final form, it may seem that some important aspects have been neglected at times. However, these apparent omissions are recognized, and pertinent manuscripts are likely in preparation or planned. The field is so very large and the interests in it are so varied that the Editor and the Editorial Board earnestly solicit authors and suggestions of underrepresented topics to make this international book series yet more useful and worthwhile.

Justification for the preparation of any review for this book series is that it deals with some aspect of the many real problems arising from the presence of foreign chemicals in our surroundings. Thus, manuscripts may encompass case studies from any country. Food additives, including pesticides, or their metabolites that may persist into human food and animal feeds are within this scope. Additionally, chemical contamination in any manner of air, water, soil, or plant or animal life is within these objectives and their purview.

Manuscripts are often contributed by invitation. However, nominations for new topics or topics in areas that are rapidly advancing are welcome. Preliminary communication with the Editor is recommended before volunteered review manuscripts are submitted.

Summerfield, North Carolina D.M.W.

Contents

The Environmental Impact of Growth-Promoting Compounds Employed by the United States Beef Cattle Industry: History, Current Knowledge, and Future Directions

Alan S. Kolok and Marlo K. Sellin

A.S. Kolok
Department of Biology, 6001 Dodge Street, University of Nebraska at Omaha, Omaha, NE, 68182-0040, USA (e-mail: akolok@mail.unomaha.edu).

M.K. Sellin
Center for Environmental Toxicology, 986805 Nebraska Medical Center, Omaha, NE, 68198-6805, USA

D.M. Whitacre (ed.), *Reviews of Environmental Contamination and Toxicology.*

1 Introduction

Recent publications in the popular literature (Renner 2002; Raloff 2002) have suggested that the growth-promoting compounds used on beef cattle in the United States may have significant impacts on local aquatic environments. Clearly, beef cattle held in finishing feedlots represent a greater environmental risk than do pasture-fed cattle for at least two principal reasons. First, cattle in feedlots are held at much higher concentrations than on pastureland. Second, most cattle in feedlots are administered high-potency growth-promoting compounds (Montgomery et al. 2001). Research has shown that these compounds can travel in the environment (Lange et al. 2002) and that they may cause endocrine-disrupting effects on local fish populations (Orlando et al. 2004).

In this review, our principal hypothesis is that the environmental risks associated with finishing beef cattle feedlots are best dealt with when there is a thorough understanding of (1) the implant strategies used at feedlots, (2) the environmental fate of the growth-promoting compounds used on the feedlot, and (3) the effect that these growth promoters have on fishes should they reach surface waters. This chapter reviews the current state of knowledge and points out areas where additional research is warranted.

2 The Beef Cattle Industry in the United States and Growth-Promoting Compounds

2.1 *Steers and Heifers*

When considering the U.S. beef cattle industry, the animal that generally comes to mind is a steer (castrated bull). Throughout history, when bulls are used for beef production, they are castrated to attenuate their aggressive tendencies. The consequence of castration is a reduction in the concentration of circulating androgens in steers (Lee et al. 1990). Castration and the loss of testis-derived androgens comes with a price, as steers experience slower rates of growth than do bulls (Jones et al. 1991; Lone 1997). To augment the growth of steers, the U.S. beef industry routinely uses exogenous growth-promoting steroids (Lone 1997).

Although steers may predominate on a cattle feedlot, the industry also makes use of heifers (young females). Heifers, tending to be smaller in mass and having a greater fat content, can also be given the same type of growth-promoting implants that are used on steers to increase their mass and percentage of lean muscle (Montgomery et al. 2001). Heifers differ from steers in that they are generally held on feedlots in an intact (not spayed) form and, unless hormonally controlled, will come into estrus (Lone 1997). Estrus causes females to display behaviors toward each other that will compromise their growth efficiency. To prevent these behaviors, estrus is inhibited using a progestogenic feed additive (Lone 1997).

2.2 *Diethylstilbestrol*

The first synthetic growth-promoting compound used on beef cattle was diethylstilbestrol (DES). Oral DES administration to beef cattle was approved in 1954, and DES implants were approved for cattle in 1955. The use of DES was found to be highly effective; for example, steers fed DES experienced a 17% increase in weight gain and a 12% increase in feed conversion to live weight (Rumsey et al. 1981). However, growing evidence (Herbst et al. 1971; Wade 1972) that DES is carcinogenic ultimately pitted advocates for the use of DES as a growth promoter in beef cattle against supporters of the Delaney Amendment to the Federal Food, Drug, and Cosmetic Act, which bans the deliberate addition of any carcinogenic compounds to food (Epstein 1990). The issue of whether DES should be banned for use by the cattle industry was debated until 1979, at which time the ban was put into full effect. The removal of DES from the market precipitated a flurry of research to evaluate other growth-promoting strategies (Montgomery et al. 2001), leading to the development of practices used by the industry today.

2.3 *Modern Growth-Promoting Compounds*

The driving force behind the development of growth-promoting compounds has been, and continues to be, optimization of return on economic investment by maximizing cattle growth, feed efficiency, and carcass quality (Montgomery et al. 2001). Currently, the U.S. Food and Drug Administration (USFDA) has approved 25 different growth-promoting implants for use in beef cattle (Montgomery et al. 2001). All these implants contain one, or at most two, of the following five growth-promoting steroids: estradiol (E_2) (in the form of 17β-E_2 or estradiol benzoate), progesterone (P), testosterone (T) propionate, trenbolone acetate (TBA), and zeranol (Parliamentary Office of Science and Technology 1999; Meyer 2001; Siemens 1996; Table 1). The USFDA has also licensed the progestogin melengestrol acetate (MGA) for use as a growth promoter for heifers. Melengestrol acetate is not administered to cattle in an implant, but rather is directly added to their feed at a dosage between 0.25 and 0.5 mg per head per day.

Two of these compounds, zeranol and T, are not generally used at finishing feedlots in large quantities but rather are used on cattle at earlier stages of their development (ZoBell et al. 2000). The reason for this is purely pragmatic; E_2 has been shown to be more anabolic than the synthetic estrogen zeranol (Siemens

Table 1 Commercially available implant formulations currently used by the beef cattle industry. See text for discussion

Compound	Quantity in implant (mg)	Potency
Single compound implants		
17β-Estradiol	24	Mild estrogen
TBA	140–200	Strong androgen
Zeranol	36–72	Mild, strong estrogen based on dose
Combination implants		
Estradiol benzoate/progesterone	10–20 100–200	Mild, strong estrogen based on dose
Estradiol benzoate/ Testosterone proprionate	20 200	Mild androgen
Estradiol[a]/TBA	8–28 40–200	Mild, strong androgen based on dose

[a]May be in the form of estradiol benzoate or 17β-estradiol depending upon the manufacturer.

1996), whereas TBA has been shown to be more anabolic, and less androgenic, than T (Preston 1997). Although current growth-promoting strategies suggest administration of low to moderate potency anabolic compounds to suckling calves or stocker cattle (cattle that are being transitioned from living on pastureland to living in a feedlot), growth efficiency strongly advocates that the highest potency growth promoters (namely TBA and E_2) be used when cattle are maintained on finishing feedlots (Siemens 1996; ZoBell et al. 2000). Given that zeranol and T are not likely to be prevalent in the waste stream emanating from finishing feedlots, they are not considered further in this review.

Cattle kept on finishing feedlots include steers, heifers, and calf-feds (large calves that are moved onto a feedlot directly following weaning). While there are similarities in the strategies used to enhance the growth of steers, heifers, and calf-feds, there are also fundamental differences. Of the high-potency implants, the most effective consistently proves to be a combination implant containing 120–200 mg TBA and 24–28 mg E_2 (Siemens 1996; Hermesmeyer et al. 2000; Foutz et al. 1997). Steers, heifers, and calf-feds are all likely to receive this implant when they are nearing the end of their tenure on the finishing feedlot (Siemens 1996). Although steers are generally only given a single implant, the strategy is different for calf-fed steers and heifers. Calf-feds are brought to the feedlot at a younger age and lighter mass than steers and therefore remain on the feedlot longer than the average steer (Montgomery et al. 2001). These longer residence times necessitate that the calf-feds receive two sets of implants. The most responsive combination of implants used on calf-fed steers is a 100 mg P:10 mg E_2, followed by the standard TBA:E_2 implant (Montgomery et al. 2001). Heifers are also given the same TBA:E_2 implant as steers, but are often concomitantly administered MGA as a food additive (Meyer 2001; Siemens 1996). Melegestrol acetate is administered to prevent estrus, thereby reducing behaviors that compromise growth (Lone 1997).

3 Metabolism and Excretion of Growth-Promoting Compounds

Administration of radiolabeled growth-promoting compounds in cattle has been used to determine routes of excretion along with the primary metabolic end products produced. Biliary excretion and fecal deposition is the primary route by which growth-promoting compounds pass through cattle into the environment. Hepatic metabolism generally begins with selective oxidation/reduction of the compounds (phase I metabolism; Table 2), followed by glucuronide or sulfate conjugation (phase II metabolism; Jones et al. 1991; Schwarzenberger et al. 1996). A brief synopsis of phase I metabolism of the four major growth-promoting compounds is discussed next.

3.1 Phase I Metabolism

3.1.1 17β-E_2

When radioactive 17β-E_2 was injected into steer calves, then followed through excretion (Ivie et al. 1986), 58% was excreted in feces and 42% was excreted in urine. The primary metabolic pathway for the excretion of 17β-estradiol is oxidation to estrone (E_1), followed by a stereoselective reduction of E_1 to 17α-estradiol (17α-E_2) (Mellin and Erb 1966a,b). Indeed, 17α-E_2 comprised more than 60% of the radiolabeled metabolites found in feces from Holstein steers injected with labeled 17β-E_2 (Ivie et al. 1986).

Table 2 The primary excreted compounds from beef cattle administered growth-promoting compounds

Administered compound	Primary excreted compounds
Estradiol[a]	17-α Estradiol, estrone
Progesterone	Progesterone (urine)
	3α-Hydroxy-5β-pregnan-20β-one (fecal)
	5β-Pregnane-3-20β-diol (fecal)
Trenbolone acetate	17α-Trenbolone, 17β-trenbolone
Melengestrol acetate	Melengestrol acetate

[a] May be in the form of estradiol benzoate or 17β-estradiol, depending upon the manufacturer.

3.1.2 TBA

The fecal route of excretion also predominates for TBA. Pottier et al. (1981) found that 80% of radiolabeled TBA was excreted in the bile 24 hr after injection. The primary metabolic pathway for the excretion of TBA is hydrolysis of the acetate to 17β-TB, oxidization to trendione, then reduction to 17α-TB (Pottier et al. 1981; Schiffer et al. 2001; Rico 1983).

3.1.3 P

As with the other compounds, the majority of P is excreted in feces (90%–96% in feces and 4%–10% in urine; Estergreen et al. 1977). The major liver metabolites of P in steers and heifers were 3α-hydroxy-5β-pregnan-20β-one and 5β-pregnane-3-20β-diol. In kidney, the major metabolites were 20β-hydroxypregan-4-en-3-one, 3α- and 3β-hydroxy-5α-pregnan-20-one, and 5α-pregnane-3β-20β-diol. Furthermore, in the kidney, 15% of the total radioactivity was parent P (Purdy et al. 1980). These results are consistent with the contention put forth by Schwarzenberger et al. (1996) that the primary metabolites of P in mammals are 5α- and 5β-reduced pregnanediones along with hydroxylated pregnanes.

3.1.4 MGA

Excretion of MGA, as with the other growth-promoting compounds, is primarily fecal. Although the liver excreted approximately 87% of the total radiolabeled MGA injected into heifers, 10%–17% of the MGA administered in the feed passed through the animal unabsorbed and in parent form (Krzeminski et al. 1981). Liver metabolism of MGA leads to the production of a number of metabolites; however, the concentration of these metabolites remains low, and the bulk of MGA appears to be excreted in parent form (Krzeminski et al. 1981).

3.2 Phase II Metabolism

Despite the fact that cattle excrete significant amounts of growth-promoting steroids in conjugated form, the environmental fate and effect of these conjugated forms are rarely discussed. It has been reported that conjugated forms of growth-promoting compounds are often deconjugated in the intestine; therefore, voided feces generally contains a higher percentage of free than conjugated growth-promoting compound (Schwarzenberger et al. 1996). Similarly, once these conjugated forms enter the environment they may be quickly deconjugated by local bacteria (D'Ascenzo et al. 2003; Belfroid et al. 1999). The current literature suggests that conjugated forms of these compounds will not persist in the environment; however, additional research is necessary to substantiate this supposition.

3.3 *Steroidogenic Activity of Feces*

Fecal pats from heifers and steers implanted with a TBA:E_2 implant have been found to contain significant estrogen- and androgen-receptor gene transcription activities but no progestogenic activity (Lorenzen et al. 2004). Levels of androgen and estrogen activity appeared to be time dependent for heifers treated solely with a TBA:E_2 implant, such that the activity in the pats decreased as the time from initial implantation increased. Interestingly, in steers receiving the combination implant, and heifers receiving the TBA:E_2 implant along with MGA in their feed, there was significant estrogenic activity detected in the fecal pats, but a lack of androgenic and progestogenic activity. The lack of progestogenic activity may not be at all surprising considering that unmetabolized progesterone is not generally detected in the fecal samples of mammals (Schwarzenberger et al. 1996).

4 Environmental Fate and Transport of Growth-Promoting Compounds

Despite the fact that growth-promoting compounds have been used by the cattle industry since the 1950s, research into the environmental fate and transport of these compounds has only been of interest to the scientific community for the past decade. Because 17β-E_2 can be released from other sources (e.g., wastewater treatment plants), the greatest amount of information exists for this compound. Given the limited amount of information available, this review borrows heavily from closely related research that has focused on other steroidogenic compounds (T), other sources of steroidogenic pollution (wastewater treatment plants), and other sources of manure-bound steroids (chickens).

4.1 *Soil and Soil Columns*

Benchtop studies investigating the fate and transport of growth-promoting compounds in soils have only been conducted on 17β-E_2 and its metabolites. *In vitro* studies featuring agricultural soils have found that much of the 17β-E_2 quickly becomes bound to soil particles. For example, Colucci et al. (2001) and Colucci and Topp (2002) applied radiolabeled 17β-E_2 and E_1 to three different agricultural soils over a range of temperatures and moistures. These experiments revealed a rapid conversion of 17β-E_2 into E_1, which was then degraded by the microbial community into a nonextractable form. Sorption experiments featuring soil–water slurries have found that 17β-E_2 quickly sorbs onto sediment particles (Lee et al. 2003; Casey et al. 2003). Furthermore, the dissipation half-lives of 17β-E_2 in soil–water slurries ranged from 0.8 to 9.7 d depending upon soil type (Lee et al. 2003). Sorption experiments featuring manure and biosolid–water slurries are consistent with the studies in which 17β-E_2 is added directly onto the soils (Jacobsen et al. 2005). Addition of swine manure slurry to the soils hastened the conversion of 17β-E_2 to E_1, as well as the mineralization of [^{14}C]-17β-E_2. This result has led to the speculation that the fate of 17β-E_2 when applied in manures is not going to be qualitatively different from that which occurs when the compound is directly applied to soils in parent form (Jacobsen et al. 2005).

All the studies just discussed agree that once 17β-E_2 or any of its metabolites interact with agricultural soil, escape of the compound is unlikely (Colucci and Topp 2002). Furthermore, it has also been suggested that any escape that does occur is likely to be associated with surface runoff and with sediment and colloidal transport (Lee et al. 2003; Casey et al. 2003).

There is little information available that focuses on the transport of 17β-E_2 through soils during irrigation. Shore et al. (1993) added labeled steroids to 1 g irrigated heavy soil. They report that 56% of the E_2 and 52% of the E_1 was washed out of the soils, while the remaining proportion was tightly bound to the soil in a non-extractable form. Unfortunately, the authors present these results as preliminary findings and do not reveal much about the experimental design associated with their results. These preliminary findings, however, are consistent with more recently collected results. Das et al. (2004) found that most (58%–93% depending upon soil type) of the 17β-E_2 used in packed soil column experiments became sorbed to sediment particles. Furthermore, the half-life of the hormone was estimated to be from a few hours to a few days. In flow-interrupted transport experiments, 17β-E_2 concentrations in the effluent decreased while E_1 concentrations increased during the flow-interruption periods. Once flow resumed and additional hormone was added to the system, the concentration of 17β-E_2 in the effluent returned to elevated levels. This result is intriguing as feedlots experience a continuous replenishing of the 17β-E_2 in the manure, and periodic rainstorms can easily release these newly deposited compounds into runoff.

4.2 Testosterone

Given that T is licensed for use as a growth-promoting compound, it is interesting to compare the fate and transport of T with 17β-E_2. Testosterone, similar to 17β-E_2, has been found to be strongly sorbed to and rapidly degraded in agricultural soils (Lee et al. 2003; Jacobsen et al. 2005; Lorenzen et al. 2005; Kim et al. 2007). Testosterone degrades more readily in agricultural soils than 17β-E_2 but does not appear to sorb to soil particles as readily as 17β-E_2 (Casey et al. 2004). During column experiments (Das et al. 2004; Casey et al. 2004), the movement of T is consistent with that of 17β-E_2; however, parent T, unlike 17β-E_2, has been found to transverse soil columns in its intact, parent form. Clearly, not all steroidogenic compounds interact with soils in the same way; therefore, assumptions regarding the fate of TBA, P, and MGA should not be made without further research.

4.3 Washout of Growth-Promoting Compounds from Agricultural Fields and Manure Piles

A very limited number of papers have been published that focus on washout of steroids or growth-promoting compounds from manure piles or agricultural fields. The studies that have been conducted on washout of 17β-E_2 from agricultural fields focused on fields fertilized with chicken litter (excreta, feathers, waste feed, bedding). The sole study that has focused on washout from beef cattle manure piles focused on TBA and MGA.

4.3.1 17β-E_2

Nichols et al. (1997) conducted a runoff study in which simulated rain was applied to pasture amended with poultry litter. They found that first-storm runoff concentrations of 17β-E_2 increased from 1.28 μg/L to 198.8 μg/L as application rates of poultry litter increased. Furthermore, during a second-storm event 7 d later, 17β-E_2 was still detected; however, losses were 66%–69% less than that from the first runoff event. The authors suggest that 17β-E_2 may persist in the field as it aggregates with chicken litter, which protects it from photo- and microbial degradation. In field trials, Finlay-Moore et al. (2000) measured concentrations of 17β-E_2 in the soils and in the runoff from fields of fescue amended with litter from broiler chickens. Before application, concentrations ranged around 55 ng/kg, whereas immediately after application concentrations increased 10 fold up to 675 ng/kg. Estradiol concentrations were significantly elevated in soils 14 d after application but returned to background levels in all plots several weeks after application. Concentrations of 17β-E_2 in the runoff from these fields increased from background levels of 50–150 to 20–2530 ng/L, depending upon litter application rate and time between application and runoff.

4.3.2 TBA and MGA

Beef cattle, in contrast to chickens, are administered androgenic, estrogenic, and progestogenic growth-promoting implants (see Table 1). Research on cattle manure has traditionally focused on the washout of TBA and MGA, while little to no information is available for 17β-E2. Schiffer et al. (2001) followed the fate of TBA and MGA administered to steers and heifers, respectively, in their dry dung, in liquid manure, and in the soil after it had been fertilized with the dry dung. The soil fertilized with solid dung contained trace amounts of trenbolone (TBOH) for 58 d and MGA for 195 d postfertilization. TBOH was also detected in solid dung for 58 d, in liquid manure for 260 d, and in soil several months after fertilization. Similarly, unmodified MGA was traceable in solid dung for 195 d and in soil several months after fertilization. Furthermore, the dung piles used in the report by Schiffer et al. (2001) released small amounts of TBOH in runoff following rainstorms.

4.4 Growth-Promoting Compounds and Steroidogenic Activity in Natural Waterways

4.4.1 17β-E_2 and Estrogenic Activity

Relationships between proximity to a feedlot and estrogen concentration (or estrogenic activity) in water have been studied with mixed results (Table 3). In a study in which 100 beef cows and calves were in close proximity to a receiving pond (5 ft away from the pond's edge), Irwin et al. (2001) detected 17β-E_2 in the pond water. Soto et al. (2004) analyzed water samples collected at six sites throughout the lower Elkhorn River (Nebraska) for estrogenic activity (E-screen), as well as for the presence of E_1 and 17β-E_2. Estrogenic activity was found at all six sites, with the highest levels of activity found in a feedlot retention basin and at confluence of the drainage ditch and the Elkhorn River (approximately 0.5 km from the retention pond). Estrone was detected at each of the six sites but accounted for little (3%–46%) of the estrogenic activity.

Similar results were obtained by Kolodziej et al. (2004). In that study, surface water samples were collected upstream and downstream of dairy farms and irrigation canal discharge points in the northeastern San Joaquin Valley of central California. Water samples were analyzed for six steroids including E_1, 17β-E_2, and estriol (E_3). Steroid hormones were only sporadically detected in the surface water samples, with no correlation being evident between steroid concentration and sampling location.

On a larger geographic scale, 17β-E_2 was detected in the water collected from five springs in northwestern Arkansas, ranging in concentration from 6 to 67 ng/L (Peterson et al. 2000). Poultry litter had been applied as a fertilizer to the recharge areas for all five springs, and the similarity in concentration profiles among 17β-E_2, fecal coliform, and *Escherichia coli* led the authors to implicate animal wastes as the source of the 17β-E_2.

In a study featuring polar organic integrative chemical samplers (POCIS), Mathiessen et al. (2006) deployed these samplers upstream and downstream from areas where livestock input was expected. One POCIS replicate was analyzed for E_1, ethynylestradiol (EE_2), and 17β-E_2, while a second replicate was analyzed for estrogenic activity using the yeast estrogen screen (YES). On 4 of the 8 farms where it was possible to make a direct comparison between upstream and downstream estrogenic activities, the downstream value was greater than the upstream value. With respect to the estrogenic metabolites, E_1 and 17β-E_2 were virtually ubiquitous. Normalized concentrations of estrogen metabolites generally agreed with estrogenic activity results as the downstream value was greater than the upstream value on 6 of 10 farms. A comparison between the YES estrogen-equivalents and the steroid analytical data from the upstream and downstream sites ($n = 22$) yielded a statistically significant result.

POCIS have also been used on the Elkhorn River to detect growth-promoting compounds (Kolok et al. 2007). POCIS were deployed at four locations for 21 d in the Elkhorn River. Two of the sites were in creeks downstream from major beef cattle feedlots, the third was downstream from the Norfolk wastewater treatment plant, and a fourth was at the same reference site used in previous studies (Orlando

Table 3 Detection of estrogenic, androgenic, and progestogenic compounds in waterways associated with runoff from animal feeding operations (see text for additional information)

Comparison	Source	Analysis	Medium	Expected* difference between locations	Frequency of detection	Reference
Estrogens/estrogenic activity:						
Onsite pond /offsite pond	Beef feedlot	17β-Estradiol	Water	Yes	E_2 detected at all sites	Irwin et al. 2001
6 different locations	Beef feedlot	Estrone, 17β-estradiol	Water	Mixed results	E_1 detected at all sites	Soto et al. 2004
6 different locations	Beef feedlot	E-Screen (estrogenic activity)	Water	Yes	Estrogenic activity detected at all sites	Soto et al. 2004
15 locations	Dairy	Estrone, 17β-estradiol, estriol	Water	No	Sporadic detection	Kolodziej et al. 2004
Upstream/downtream (8 comparisons)	Various (dairy, beef, pigs)	Yeast estrogen screen (estrogenic activity)	POCIS	Yes (4 of 8 streams)	Estrogenic activity at all sites	Matthiessen et al. 2006
Upstream/downstream (10 comparisons)	Various (dairy, beef, pigs)	Estrone, 17β-estradiol	POCIS	Yes (6 of 10 streams)	E_1 detected at all sites	Matthiessen et al. 2006
4 locations	Beef feedlot	Estrone, 17β-estradiol	POCIS	Yes (reference vs. 3 sites)	E_1 detected at all sites	Kolok et al. 2007
Androgens/androgenic activity:						
6 different locations	Beef feedlot	17α-Trenbolone, 17β-trenbolone, trendione	Water	No	Marginal levels only	Soto et al. 2004
6 different locations	Beef feedlot	A-Screen (androgenic activity)	Water	Yes	Androgenic activity at all sites	Soto et al. 2004
15 locations	Dairy	Androstenedione, testosterone	Water	No	Rarely detected	Kolodziej et al. 2004
Upstream/drain/downstream	Beef feedlot	17α-Trenbolone, 17β-trenbolone	Water	Yes	Generally found in drain only	Matthiessen et al. 2006
Upstream/drain/downstream	Beef feedlot	Androgenic activity	Water	Yes	Highest activities generally in drain	Matthiessen et al. 2006

(continued)

Table 3 (continued)

Comparison	Source	Analysis	Medium	Expected* difference between locations	Frequency of detection	Reference
6 locations	Beef feedlot	17β-Trenbolone	POCIS	No	Not detected at any site	Kolok et al. 2007
Progestogins/MGA:						
15 locations	Dairy	Progesterone, medroxy-progesterone	Water	No	P not detected at any site. MP <1 ng/L at all sites	Kolodziej et al. 2004
6 locations	Beef feedlot	Progesterone, meleges-trol acetate	POCIS	No	P and MGA detected at all locations	Kolok et al. 2007

POCIS, polar organic integrative chemical samplers.
* Did results support a priori expectations?

et al. 2004; Soto et al. 2004). POCIS were analyzed for E_1, 17β-E_2, and E_3, and as with the results from Mathiessen (2006), E_1 (but not 17β-E_2) was detected at all six locations. No clear-cut relationship was evident between the concentration of E_1 and proximity to cattle feedlot operations.

4.4.2 TBA and Androgenic Activity

Where relationships between proximity to a feedlot and androgen concentration (or androgenic activity) in water have been determined, the results are no more compelling than those found for estrogenic compounds (see Table 3). Soto et al. (2004) measured TBA metabolites and androgenic activity at six locations in the Elkhorn River watershed. The highest level of androgenic activity was found in a feedlot retention pond. The other four (nonreference) field sites had approximately 40% of the activity found in the retention pond, whereas the reference site had approximately 20% of the activity found in the retention pond. Water from these sites had only marginal levels of TBA metabolites, leading the authors to suggest that their results may have been artifactual and that the observed androgenic activity may have been attributed to natural androgens. A follow-up study in the Lower Elkhorn featuring POCIS deployed in the field for 21 d at six locations did not detect 17β-TB in two small creeks that drained watersheds containing major feedlot operations (Kolok et al. 2007).

Durhan et al. (2006) also determined the concentration of 17α-TB and 17β-TB in a river receiving discharge from a beef cattle feedlot. They sampled river water at nine different times, over the course of 2 yr, above, below, and at the discharge structure. The occurrence and magnitude of the TB metabolites varied considerably from one sampling period to the next, with no clear-cut seasonal pattern. Furthermore, although the presence of TBA metabolites tended to be highest at the discharge, this did not hold true for every sampling period.

4.4.3 P and MGA

There are only two studies of which we are aware that have looked for P and MGA in waterways downstream from cattle feeding operations. One (Kolodziej et al. 2004) looked for P and medroxyprogesterone downstream from dairy farms in the San Joaquin Valley of central California. Progesterone was not collected at any of the six field sites, whereas medroxyprogesterone was detected at trace levels (less than 1 ng/L). The second study (Kolok et al. 2007) deployed POCIS at six locations in the Elkhorn River, Nebraska, for 21 d, then looked for MGA and P in POCIS extracts. Both P and MGA were detected at all six sites, despite the fact that at least two of the sites were not in direct proximity to beef cattle feedlots.

5 Biological Effects of Growth-Promoting Compounds on Fishes

As was true for fate and transport, information on the biological effects of cattle growth-promoting compounds to fish and wildlife is quite extensive in some areas but virtually nonexistent in others. Fish have been the aquatic organisms of choice for study because of their commercial importance and their history of use as environmental sentinel organisms. As such, this review focuses on the literature on fish exposed to growth-promoting compounds. Only one published study has focused on the effects of growth-promoting compounds in the field, but numerous studies have demonstrated that exposure to growth-promoting compounds affects the endocrine and reproductive systems of fishes. These studies are pharmacological in nature (i.e., fish are exposed to a parent compound rather than being exposed to the compound within a biological matrix) and focus on a variety of endpoints that span from molecular to the whole organism. For all endpoints, the primary emphasis has been on 17β-E_2 and, to a lesser extent, TBA metabolites.

Research on the biological effects of growth-promoting compounds appears to be focused on two ultimate goals: (1) identification of molecular/cellular/tissue level biomarkers of exposure that can be used to signify when fish have been exposed to growth-promoting compounds, and (2) demonstration of reduced reproductive fitness as a consequence of exposure to growth-promoting compounds. Altered sex steroids, gene expression, and plasma vitellogenin (vtg) concentrations are all molecular endpoints of exposure, whereas alterations in secondary sexual characteristics and gonad health can be evaluated as tissue-level endpoints of exposure. With respect to reproductive fitness, impaired reproduction and skewed sex ratios resulting from altered sexual differentiation are the only endpoints that have been measured to date. With very few exceptions, no effort has been put into determining the fitness consequences of alterations in the more reductionistic endpoints.

5.1 Sex Steroids, Vitellogenin Production, and Gene Expression Patterns

5.1.1 Sex Steroids

Exposure to exogenous growth-promoting compounds has been shown to alter circulating levels of endogenous sex steroids. Because one sex steroid, for example, T, acts as an intermediate metabolite for the production of other sex steroids (11-kT, 17b-E2), the effects of exposure to growth-promoting compounds are not easily characterized.

17β-E_2

Exposures to 17β-E_2 have been shown to alter circulating sex steroid concentrations and sex steroid production in fishes. Kramer et al. (1998) found that exposures to

17β-E_2 lead to dose-dependent increases in plasma 17β-E_2 concentrations in male fathead minnows (*Pimephales promelas*). Androgen production is also responsive to 17β-E_2. Loomis and Thomas (2000) examined the effect of 17β-E_2 on 11-ketotestosterone (11-KT, the primary fish androgen) production using an *in vitro* incubation bioassay system featuring testicular tissue from Atlantic croaker (*Micropogonias undulates*). It was determined that 17β-E_2 significantly decreases 11-KT synthesis at concentrations greater than 367 nM E_2.

TBA

In female fathead minnows exposed to 17β-TB, concentrations of plasma T and 17β-E_2 exhibited a U-shaped dose–response relationship (Ankley et al. 2003). Specifically, females exposed to intermediate concentrations (50 and 500 ng/L) experienced significant reductions in both sex steroids; however, at higher concentrations (50,000 ng/L), plasma steroid concentrations appeared to be somewhat restored. Male sex steroid concentrations were also altered by exposures to 17β-TB. At 50,000 ng/L, males experienced a significant increase in plasma 17β-E_2 concentrations and a significant decrease in plasma 11-KT concentrations relative to controls. Similarly, Jensen et al. (2006) showed that females exposed to 17α-TB experienced significant reductions in plasma T and 17β-E_2 concentrations. However, 17α-TB did not alter male sex steroid concentrations at the exposure concentrations tested.

P

To our knowledge, only one study has examined the effects of P on sex steroid synthesis in fishes, and this study is somewhat limited in that only 11-kT production was assessed. The findings of this study, using an *in vitro* bioassay featuring Atlantic croaker testicular tissue, did not find that P altered 11-kT synthesis (Loomis and Thomas 2000).

Vitellogenin Production

Vitellogenin is an egg precursor protein produced in female fishes in response to endogenous estrogen production. Exposures to 17β-E2 can lead to inappropriately elevated vtg concentrations in males, and exposure to TBA can lead to inappropriately low levels of vtg in females.

17β-E_2

Multiple studies have demonstrated that 17β-E_2 exposures lead to vtg production in male fishes (Table 4). For example, Panter et al. (1998) showed that male fathead minnows exposed to concentrations of 17β-E_2 of 100–1000 ng/L experience a significant increase in plasma vtg concentrations relative to controls. Similar results

Table 4 Summary of the effects of growth promoters and their metabolites on vitellogenesis in fishes

Compound	Species	Observed effect on vitellogenin (vtg) concentration	Reference
17β-Estradiol	Fathead minnow	Increased male plasma vtg	Panter et al. 1998
		Increased male and female plasma vtg	Seki et al. 2006
	Rainbow trout	Increased plasma vtg among juveniles	Thorpe et al. 2003
	Zebrafish	Increased whole-body vtg among juveniles	Brion et al. 2004
		Increased whole-body vtg among males and females	Brion et al. 2004
		Increased male and female plasma vtg	Seki et al. 2006
	Medaka	Increased male and female plasma vtg	Seki et al. 2006
Estrone	Fathead minnow	Increased male plasma vtg	Panter et al. 1998
	Zebrafish	Increased whole-body vtg among juveniles	Holbech et al. 2006
	Rainbow trout	Increased plasma vtg among juveniles	Thorpe et al. 2003
Estriol	Zebrafish	Increased whole-body vtg among juveniles	Holbech et al. 2006
17β-Trenbolone	Fathead minnow	Decreased female plasma vtg	Ankley et al. 2003; Seki et al. 2006
	Medaka	Decreased female plasma vtg	Seki et al. 2006
		Decreased whole-body vtg among juveniles	Orn et al. 2006
	Zebrafish	Decreased female plasma vtg	Seki et al. 2006
		Decreased whole-body vtg among juveniles	Holbech et al. 2006; Orn et al. 2006
17α-Trenbolone	Fathead minnow	Decreased female plasma vtg	Jensen et al. 2006

have been reported for several species including cunner (*Tautogolabrus adspersus*; Mills et al. 2003), rainbow trout (*Onchorynchus mykiss*; Thorpe et al. 2003), zebrafish (*Danio rerio*; Brion et al. 2004), and stickleback (*Gasterosteus aculeatus*; Hahlbeck et al. 2004). Estrogen metabolites, including E_1 and E_3, have also been shown to induce male vtg production in a variety of fish species.

TBA

Several studies have clearly shown that exposures to TBA metabolites alter female plasma vtg concentrations (see Table 4). Reductions in female plasma vtg have been documented in medaka (*Oryzias latipes*) exposed to 17β-TB concentrations greater than 39.7 ng/L (Seki et al. 2006; Orn et al. 2006), in zebrafish exposed to concentrations greater than 50 ng/L (Hohbech et al. 2006; Orn et al. 2006; Seki et al. 2006), and in fathead minnows exposed to greater than 50 ng/L (Ankley et al. 2003, 2004; Seki et al. 2006). Similarly, exposures to 30 and 100 ng/L 17α-TB also reduce female vtg concentrations (Jensen et al. 2006). Reductions in female vtg are one of the few examples in which alterations in a molecular biomarker have been shown to be correlated with reproductive fitness, as Miller et al. (2007) found a strong correlation between female vtg concentrations and fecundity. Whether this correlation is indicative of a cause-and-effect relationship or is associated with another endpoint that is driving the relationship remains to be demonstrated.

In addition to the reduction in vtg in females, one study has shown alterations in vtg following exposures to 17β-TB. Ankley et al. (2003) found that exposure to 50,000 ng/L 17β-TB induces male vtg production. Evidence suggests that these alterations in both female and male vtg are linked to alterations in circulating concentrations of 17β-E_2 (Ankley et al. 2003).

P

Pelissero et al. (1993) evaluated the ability of P to induce vtg synthesis in cultured rainbow trout hepatocytes from immature females and found that P did induce vtg secretion. However, the potency of P was nearly 1000 times less than that of E_2. *In vivo* studies using hyophysectomized female catfish (*Heteropneustes fossilis*) have found that P does not alter plasma vtg concentrations (Sundararaj and Panchanan 1981).

MGA

Vitellogenin gene expression in rainbow trout hepatocyte cultures was not induced by MGA, suggesting that it does not affect vitellogenesis (LeGuevel and Pakdel 2001).

5.1.2 Gene Expression Patterns

Differential gene expression is a powerful tool that can be used to elucidate the mechanisms by which growth-promoting compounds lead to their ultimate effects. Currently, a lack of knowledge regarding the function of the differentially expressed genes limits the explanatory power of this approach.

17β-E_2

Several studies have conducted gene expression analysis on fishes exposed to E_2. These studies consistently find that 17β-E_2 exposures lead to an increase the expression of genes associated with egg production, such as vtg and choriogenin (Arukwe et al. 2001; Larkin et al. 2002; Moens et al. 2006; Table 5). Hepatic

Table 5 Summary of the effects of growth promoters and their metabolites on hepatic gene expression in fishes

Compound	Species	Observed effect	Reference
17β-Estradiol	Rainbow trout	Increased vitellogenin expression Increased choriogenin expression	Arukwe et al. 2001
	Largemouth bass	Increased vitellogenin expression Increased choriogenin expression Increased aldose reductase expression Increased aspartic protease expression	Larkin et al. 2002
	Fathead minnow	Increased estrogen receptor-α expression	Filby and Tyler 2005
	Common carp	Increased vitellogenin expression Increased choriogenin expression Increased glutathione S-transferase expression Increased calmodulin expression Increased intelectin 2 expression Increased myeloid protein-1 expression Increased atrial natriuretic peptide receptor B precursor expression Increased ribosomal protein S15 expression Decreased β-tubulin expression Decreased fetuin (long form) expression Decreased apolipoprotein B expression	Moens et al. 2006
17β-Trenbolone	Mosquitofish	Increased androgen receptor-α and -β expression[a]	Sone et al. 2005
	Fathead minnow	Decreased vitellogenin expression	Ankley et al. 2004; Miracle et al. 2006
	Rainbow trout	Increased apolipoprotein A-1 expression Increased NADH dehydrogenase expression Increased cytochrome oxidase expression Increased programmed cell death 5 expression	Hook et al. 2006

[a] Expression was not measured in the liver, but rather in the anal fin.

expression of estrogen receptor-α (ERα) has also been shown to increase in response to 17β-E_2. Specifically, male fathead minnows (Filby and Tyler 2005) and juvenile rainbow trout (Arukwe et al. 2001) exposed to 17β-E_2 experience significant increases in hepatic ER α expression. In addition to theses genes, microarray technology has identified a multitude of genes that appear to undergo changes in expression patterns following 17β-E_2 exposures. Using the common carp, Moens et al. (2006) identified 35 hepatic genes that exhibited altered expression following exposure to 17β-E_2. For some of these genes, such as vtg, there may be a link between altered expression and higher order outcomes including altered reproductive fitness. For most of the genes, however, the biological consequences of altered expression are much less clear.

TBA

Studies have also investigated the effects of 17β-TB on gene expression patterns in fish (see Table 5). Reductions in hepatic vtg gene expression have been demonstrated in females exposed to 17β-TB (Ankley et al. 2004; Miracle et al. 2006). 17β-TB also alters the expression of androgen receptor (AR) genes. Specifically, 17β-TB-exposed female mosquitofish (*Gambusia affinis*) experience significant increases in the expression of ARα and ARβ in their anal fin (Sone et al. 2005). Hook et al. (2006) used DNA microarrays to assess gene expression in rainbow trout exposed to 17β-TB and found alterations in the expression of 64 genes. Although many (nearly 50%) of the altered genes were classified as having an unknown function, several of the altered genes with known functions were identified as playing a role in transport, protein binding, enzyme activity, ion binding, and immune response. As is the case for 17β-E_2, research is needed to link TBA-induced alterations in gene expression to higher-order outcomes.

5.2 *Secondary Sexual Characteristics and Gonad Health*

Given the steroidal nature of the growth-promoting compounds, it is not at all surprising that they exhibit pronounced effects on gonad health, gonad maturation, and secondary sexual characteristics.

5.2.1 Gonad Health and Maturation

Effects on gonadosomatic index (GSI) and histological endpoints on cell types within the testes and ovary are difficult to generalize. Nevertheless, it seems likely that alteration in gonadal germ cells would have direct fitness consequences to the fish, and as such may lead to a linkage between metabolic and population outcomes.

17β-E_2

Exposures to 17β-E_2 and its metabolites have been shown to alter the gonads of fish. Panter et al. (1998) found that male fathead minnows exposed to 320 or 1000 ng/L 17β-E_2 or 317.7 ng/L E_1 experienced significant reductions in gonadosomatic index relative to controls indicating an inhibition of testicular growth. In addition, Miles-Richardson et al. (1999) found that the testes of 17β-E_2-exposed males contained hyperplastic and hypertrophied Sertoli cells and degenerate spermatozoa. The increase in Sertoli cell proliferation leads to either partial or total occlusion of lumina of the seminiferous tubules in males exposed to 17β-E_2. Histological alterations in the ovaries of 17β-E_2-exposed females were also noted. Specifically, the ovaries of 17β-E_2-exposed females contained a significantly higher proportion of secondary follicles than ovaries from control females. Furthermore, 17β-E_2 exposures reduced the proportion of mature follicles and increased the proportion of atretic follicles. These findings suggest that 17β-E_2 adversely affects the maturation of ovarian follicles.

TBA

Studies have shown alterations in the gonads of both male and female fish exposed to 17β-TB. In a study by Ankley et al. (2003), the ovaries of female fathead minnows exposed to 50 and 500 ng/L 17β-TB had a mean stage of 3.75 whereas the ovaries of control females had a mean stage of 4.5, indicating that 17β-TB slows follicle maturation. Furthermore, preovulatory atretic follicles were found in most (90%) of the ovaries from females exposed to 17β-TB. Alterations in the gonad morphology of 17β-TB-exposed male minnows were also detected. Specifically, the testis of minnows exposed to 17β-TB exhibited thinned germinal epithelium and expanded, sperm-filled lumen, indicating that 17β-TB leads to the hyperproduction of sperm. Orn et al. (2006) found that exposures to 17β-TB significantly increased the testicular area of zebrafish and the spermatozoa percentage of both zebrafish and medaka, suggesting that 17β-TB accelerates gonadal differentiation and stimulates spermatogenesis.

P

Studies have suggested that progestogins (i.e., 17α-hydroxy progesterone and 17α, 20β-dihydroxy-4-pregnen-3-one) play a significant role in oocyte maturation (Nayak et al. 2001) and spermatogenesis (Miura et al. 2006); therefore, it is not unlikely that waterborne exposures to P, and possibly MGA, would have a significant effect on gonad function.

Secondary Sexual Characteristics

The primary effects of growth-promoting compounds on secondary sexual characteristics are the demasculinization and feminization of male fish exposed to 17β-E2 and the masculinization of female fish exposed to TBA metabolites. To date, no studies have presented alterations of secondary sexual characteristics by P or MGA.

17β-E_2

Miles-Richardson et al. (1999) found that exposures to 17β-E_2 reduce the prominence of male secondary sexual characteristics in fathead minnows (Table 6). Specifically, male minnows exposed to 2 and 10 nM 17β-E_2 experience significant reductions in the diameter of breeding tubercles and in fatpad size. Studies have also shown that 17β-E_2 exposures can lead to the development of female secondary sexual characteristics in male fish. For example, exposures to 17β-E_2 lead to the development of urogenital papillae in both male zebrafish (Brion et al. 2004) and male medaka (Balch et al. 2004).

Table 6 Summary of the effects of growth promoters and their metabolites on the secondary sexual characteristics of fishes

Compound	Species	Observed effect	Reference
17β-Estradiol	Fathead minnow	Reduced tubercle diameter and fat-pad size in males	Miles-Richardson et al. 1999
	Medaka	Development of urogenital papillae in males	Balch et al. 2004
	Zebrafish	Development of urogenital papillae in males	Brion et al. 2004
17β-Trenbolone	Fathead minnow	Development of breeding tubercles in females	Ankley et al. 2003
		Development of dorsal pad in females	Ankley et al. 2003
		Development of breeding tubercles in females	Ankley et al. 2003; Seki et al. 2006
	Mosquitofish	Development of gonopodium in females	Sone et al. 2005
	Medaka	Development of papillary processes in females	Seki et al. 2006
17α-Trenbolone	Fathead minnow	Development of breeding tubercles in females	Jensen et al. 2006

TBA

Several studies have shown that exposures to 17β-TB lead to the expression of male secondary sexual characteristics in female fish (Table 6). For example, female fathead minnows develop breeding tubercles consistent with those of breeding males following exposures to both 17α- and 17β-TB (Ankley et al. 2003, 2004; Seki et al. 2006). The development of male secondary sexual characteristics has also been documented in female mosquitofish (Sone et al. 2005) and medaka (Seki et al. 2006) exposed to 17β-TB. The development of male secondary sexual characteristics in female fish exposed to 17β-TB appears to be mediated via the AR. Ankley et al. (2004) demonstrated that the development of breeding tubercles in female fathead minnows exposed to 17β-TB can be blocked with flutamide, an AR-antagonist, indicating that the development of breeding tubercles is caused by the binding of 17β-TB to the AR.

5.3 *Reproduction and Alteration in Sex Ratio*

Exposures to growth-promoting compounds in fishes may lead to a number of outcomes; however, if the exposure does not lead to a population-wide effect (extirpation of local populations), then the environmental importance of the impact can certainly be called into question. Two outcomes of exposure that can have direct fitness consequences are reduced reproductive success and dramatic alteration in sex ratio.

5.3.1 Reproduction

17β-E_2

Exposures to 17β-E_2 have been shown to alter the reproductive success of fishes. Kramer et al. (1998) found that fathead minnows exposed to 6.6 and 120 ng/L experienced a 10% and 50% reduction, respectively, in the number of eggs produced. Similar results have also been obtained for Japanese medaka (Shioda and Wakabayashi 2002; Imai et al. 2005), Java medaka (*Oryzias javanicus*; Imai et al. 2005), and zebrafish (Brion et al. 2004). A study by Imai et al. (2007) also determined that E_1 adversely effects reproduction, as Java medaka exposed to 1188 and 3701 ng/L E_1 had a significant reduction in fertility and the number of eggs produced relative to controls. Brion et al. (2004) concluded that reductions in the fecundity of zebrafish exposed to 100 ng/L 17β-E_2 were attributable to reductions in spawning frequency, suggesting that 17β-E_2 exposures may cause a disruption in mating behavior.

TBA

Exposures to TBA metabolites also adversely affect the reproduction of fishes. Specifically, fathead minnows exposed to concentrations of 17β-TB greater than 50 ng/L experience significant reductions in fecundity (Ankley et al. 2003). Similarly, fathead minnows exposed to 17α-TB experienced a significant reduction in fecundity (Jensen et al. 2006).

5.3.2 Alterations in Sex Ratio

Exposures to endocrine-active compounds during early development can skew the sex ratio of a population producing either male- or female-predominate populations (Table 7). This outcome has, in fact, been known by aquaculturists for at least 10 yr, as they have routinely used these compounds to produce unisex (i.e., nonreproductive) populations of fish for stocking. In some cases, as noted below, these exposures can also result in the production of intersex individuals.

17β-E_2

Exposures to 17β-E_2 during early development have been show to alter the sex ratio of exposed populations (Table 7). Arslan and Phelps (2004a) were able to produce predominately female (99.3%) populations of bluegill by feeding fry a diet containing 200 mg/kg 17β-E_2 for 45 d. In addition to oral administration, periodic immersions and continuous waterborne exposures to 17β-E_2 lead to female-biased sex ratios. For example, periodic immersions in a 1 mg/L 17β-E_2 solution produced 76.9% female

Table 7 Summary of the effects of growth promoters and their metabolites on sexual differentiation in fishes

Compound	Species	Observed effect	Reference
17β-Estradiol	Bluegill	Female-biased sex ratios after oral administration for 45 d	Arslan and Phelps 2004a
		Female-biased sex ratios following periodic immersions	
	Zebrafish	Female-biased sex ratios	Holbech et al. 2006
	Medaka	Female-biased sex ratios	Balch et al. 2004
		Development of testis-ova	Hartley et al. 1998
		Development of ovarian tissue in males	Hirai et al. 2006
Estrone	Zebrafish	Female-biased sex ratios	Holbech et al. 2006
Estriol	Zebrafish	Female-biased sex ratios	Holbech et al. 2006
Trenbolone acetate	Channel catfish	Male-biased sex ratios following oral administration	Galvez et al. 1995
	Blue tilapia	Male-biased sex ratios following oral administration	Galvez et al. 1996
	Bluegill	Development of intersex individuals following oral administrations	Al-Ablani and Phelps 2002
	Black crappie	Male-biased sex ratios after periodic immersions	Arslan and Phelps 2004b
17β-Trenbolone	Zebrafish	Male-biased sex ratios	Orn et al. 2006; Holbech et al. 2006

Except where mentioned, all exposures were waterborne and continuous.

bluegill (*Lepomis macrochirus*) populations. Continuous exposures to 1,000 ng/L 17β-E_2 during early development led to the production of 98% and 100% female populations of wild and female leukophore-free strains of Japanese medaka, respectively (Balch et al. 2004). In addition to feminizing populations of fishes, 17β-E_2 exposures have also been shown to lead to the development of intersex individuals. For example, more than 50% of Japanese medaka fry exposed to 4,000 and 29,400 ng/L 17β-E_2 had testis-ova whereas none of the control medaka exhibited testis-ova (Hartley et al. 1998). Similarly, 100% of genetic male medaka exposed to 140.6 ng/L 17β-E_2 had ovarian tissue whereas only 6.8% of males exposed to 33.5 ng/L 17β-E_2 had ovarian tissue (Hirai et al. 2006). Estradiol metabolites have also been shown to disrupt sexual differentiation. Holbech et al. (2006) found that exposures to 50 and 98 ng/L E_1 or 21,700 ng/L E_3 produced zebrafish populations with a significantly higher proportion of females relative to controls. Exposures to 17β-E_2 and its metabolites during early development pose an obvious threat to exposed populations because they are capable of producing female-biased sex ratios and intersex individuals.

TBA

Exposures to TBA and its metabolites have been shown to produce male-biased fish populations (Table 7). For example, oral administration of TBA has been used by aquaculturists to produce predominately male populations of channel catfish (*Ictalurus punctatus*; Galvez et al. 1995) and blue tilapia (*Oreochromis aureus*; Galvez et al. 1996). Oral administration of TBA has also been shown to produce intersex fish. Al-Ablani and Phelps (2002) found that populations of bluegill fed TBA-supplemented diets consisted of between 40% and 69% intersex individuals (depending upon exposure dose and duration) whereas control populations did not contain any intersex fish. Waterborne exposures to TBA have also been shown to alter the sex ratios of exposed populations. Arslan and Phelps (2004) were able to create all-male populations of black crappie (*Pomoxis nigromaculatus*) by periodic immersions in 1 mg/L TBA. Orn et al. (2006) found that continuous exposure to 50 ng/L 17β-TB during early development resulted in an all-male population of zebrafish. When the results of these studies are considered together, it is clear that TBA exposures have the potential to disrupt sexual development, leading to male-biased sex ratios and an increase in the incidence of intersex individuals.

6 Field Studies

To date, there is only one published study that has focused on the effect of feedlot runoff on a wild fish species (Orlando et al. 2004). In that study, adult fathead minnows were collected from three sites: a reference site, an intermediate site (no cattle feedlots in the immediate vicinity), and a contaminated site at the confluence between a drainage ditch (draining a major feedlot) and the Elkhorn River, Nebraska. A number of physiological effects were noted in the fish among the three sites. Males from the contaminated and intermediate sites had significantly smaller testes, diminished secondary sexual characteristics, and a reduction in T synthesis (from cultured testicular tissue) relative to fish from the reference site. In females, ovarian synthesis of 17β-E_2 and T was different among the three sites such that the estrogen:androgen ratio was significantly decreased at the intermediate and contaminated sites relative to the reference site.

It is not trivial to note that most of the physiological endpoints measured in Orlando et al. (2004) are not the same endpoints discussed in the previous section of this review. This field research was conducted in 1999, before much of the laboratory research presented in this review was conducted. It would be valuable to validate the results of Orlando et al. (2004) using the widely accepted endpoints discussed above.

7 Summary

The current state of knowledge regarding the environmental impact of growth-promoting compounds associated with the U.S. beef cattle industry is extensive in some areas but virtually nonexistent in others. The compounds administered to the cattle are quite well understood, as are bovine metabolism and excretion. If the sex and age of the cattle on the feedlot are known, the metabolites excreted by the cattle should be predictable with a great deal of accuracy. The fate, transport, and biological effects of growth-promoting compounds are just beginning to be studied. Most of the research conducted on the fate and transport of growth-promoting compounds has focused on 17β-E_2; however, much of this research was not conducted using feedlot runoff or manure. Studies are needed that focus specifically on manures and runoff from experimental or commercial feedlots. To date, the degree to which growth-promoting compounds are released from feedlots in a bioavailable form remains a point of speculation.

The environmental fate and transport of TBA, P, and MGA have not been well studied. Comparisons between the fate and transport of T and 17β-E_2, however, make it clear that compounds with similar structure may behave very differently once released into the environment.

Considering that 17β-E_2 is a naturally occurring estrogen and that TBA is a nonaromatizable androgen, it is not surprising that these compounds directly impact the reproductive physiology of fishes. The effects of these two compounds have been well documented, as has been described here; however, the effects of P and MGA exposures have gone largely uninvestigated. This is a serious critical gap in our knowledge base because progestogins play an important role in sex steroid synthesis and reproduction. Clearly, additional research on the consequences of exposures to P and MGA is warranted.

The majority of research investigating the effects of 17β-E_2 and TBA metabolites on fish has been conducted in the laboratory and has typically focused on continuous, pharmacological exposures to single compounds. These exposures may not bear much similarity to environmentally relevant exposures, and as such may offer little information regarding biological effects seen in nature. Cattle feedlot runoff is likely to contain a suite of growth-promoting compounds rather than any single compound. Clearly, deciphering the biological effects of exposure to complex mixtures containing androgenic, estrogenic, and progestogenic compounds will remain an important area of study for the next few years.

A second complexity associated with the biological runoff from cattle feedlots is the discontinuous nature of the release. It is likely that inadvertent entry of growth-promoting compounds will follow spring snowmelt or rainstorm events. These events will result in intermittent, pulsed exposures to high concentrations of these compounds interspersed by long-term exposures to lower concentrations. The effects of exposure timing and duration should be considered to generate a clearer understanding of the biological consequences of exposures to growth-promoting compounds.

To date, a very limited number of studies (only one!) have sought to determine whether fish living in waterways receiving runoff from cattle feedlots are adversely affected by growth-promoting compounds associated with the runoff. Clearly, more field studies need to be conducted before a relationship between cattle feedlot effluent and biological consequences can be elucidated.

Acknowledgments The authors acknowledge partial financial support from the Nebraska Department of Environmental Quality, and the University of Nebraska, Institute of Agriculture and Natural Resources, Agricultural Research Division.

References

Al-Ablani SA, Phelps RP (2002) Paradoxes in exogenous androgen treatments of bluegill. J Appl Ichthyol 18:61–64.

Ankley GT, Jensen KM, Makynen EA, Kahl MD, Korte JJ, Hornung MW, Henry TR, Denny JS, Leino RL, Wilson VS, Cardon MC, Hartig PC, Gray LE (2003) Effects of the androgenic growth promoter 17-β-trenbolone on fecundity and reproductive endocrinology of the fathead minnow. Environ Toxicol Chem 22:1350–1360.

Ankley GT, Defoe DL, Kahl MD, Jensen KM, Makynen EA, Miracle A, Hartig P, Gray LE Cardon M, Wilson V (2004) Evaluation of the model anti-androgen flutamide for assessing the mechanistic basis of responses to an androgen in the fathead minnow (*Pimephales promelas*). Environ Sci Technol 38:6322–6327.

Arslan T, Phelps RP (2004a) Directing gonadal differentiation in bluegill, *Lepomis macrochirus* (Rafinesque), and black crappie, *Pomoxis nigromaculatus* (Lesueur), by periodic estradiol-17β immersions. Aquac Res 35:397–402.

Arslan T, Phelps RP (2004b) Production of monosex male black crappie, *Pomoxis nigromaculatus*, populations by multiple androgen immersion. Aquaculture 234:561–573.

Arukwe A, Kullman SW, Hinton DE (2001) Differential biomarker gene and protein expressions in nonylphenol and estradiol-17β treated juvenile rainbow trout (*Oncorhynchus mykiss*). Comp Biochem Physiol 129C:1–10.

Balch GC, Shami K, Wilson PJ, Wakamatsu Y, Metcalfe CD (2004) Feminization of female leukophore-free strain of Japanese medaka (*Oryzias latipes*) exposed to 17β-estradiol. Environ Toxicol Chem 23:2763–2768.

Belfroid AC, Van der Horst A, Vathaak AD, Schafer AJ, Rijs GBJ, Wegener J, Cofino WP (1999) Analysis and occurrence of estrogenic hormones and their glucuronides in surface water and waste water in The Netherlands. Sci Total Environ 225:101–108.

Brion F, Tyler CR, Palazzi X, Laillet B, Porcher JM, Garric J, Flammarion P (2004) Impacts of 17β-estradiol, including environmentally relevant concentrations, on reproduction after exposure during embryo-larval-, juvenile- and adult-life stages in zebrafish (*Danio rerio*). Aquat Toxicol 68:193–217.

Casey FXM, Larsen GL, Hakk H, Simunek J (2003) Fate and transport of 17β-estradiol in soil-water systems. Environ Sci Technol 37:2400–2409.

Casey FXM, Hakk H, Simunek J, Larsen GL (2004) Fate and transport of testosterone in agricultural soils. Environ Sci Technol 38:790–798.

Colucci MS, Topp E (2002) Dissipation of part-per-trillion concentrations of estrogenic hormones from agricultural soils. Can J Soil Sci 82:335–340.

Colucci MS, Bork H, Topp E (2001) Persistence of estrogenic hormones in agricultural soils: I. 17β-Estradiol and estrone. J Environ Qual 30:2070–2076.

Das BS, Lee LS, Rao PSC, Hultgren RP (2004) Sorption and degradation of steroid hormones in soils during transport: column studies and model evaluation. Environ Sci Technol 38:1460–1470.

D'Ascenzo G, Di Corcia A, Gentili A, Mancini R, Mastropasqua R, Nazzari M, Samperi R (2003) Fate of natural estrogen conjugates in municipal sewage transport and treatment facilities. Sci Total Environ 302:199–209.

Durhan EJ, Lambright CS, Makynen EA, Lazorchak J, Hartig PC, Wilson VS, Gray LE, Ankley GT (2006) Identification of metabolites of trenbolone acetate in androgenic runoff from a beef feedlot. Environ Health Perspect 114(suppl 1):65–68.

Epstein SS (1990) The chemical jungle: today's beef industry. Int J Health Sci 20:277–280.

Estergreen VL, Lin MT, Martin EL, Moss GE, Branen AL, Luedecke LO, Shimoda W (1977) Distribution of progesterone and its metabolites in cattle tissues following administration of progesterone-4-C1,2,3,4. J Anim Sci 46:642–651.

Filby AL, Tyler CR (2005) Molecular characterization of estrogen receptors 1, 2a, and 2b and their tissue and ontogenic expression profiles in fathead minnow (*Pimephales promelas*). Biol Reprod 73:648–662.

Finlay-Moore O, Hartel PG, Cabrera ML (2000) 17β-Estradiol and testosterone in soil and runoff from grasslands amended with broiler litter. J Environ Qual 29:1604–1611.

Foutz CP, Dolezal HG, Gardner TL, Gill DR, Hensley JL, Morgan JB (1997) Anabolic implant effects on steer performance, carcass traits, subprimal yields, and longissimus muscle properties. J Anim Sci 75:1256–1265.

Galvez JI, Mazik PM, Phelps RP, Mulvaney DR (1995) Masculinization of channel catfish *Ictalurus punctatus* by oral administration of trenbolone acetate. J World Aquat Sci 26:378–383.

Galvez JI, Morrison JR, Phelps RP (1996) Efficacy of trenbolone acetate in sex inversion of the blue tilapia *Oreochromis aureus*. J World Aquac Soc 27:483–486.

Hahlbeck E, Katsiadaki I, Mayer I, Adolfsson-Erici M, James J, Bengtsson BE (2004) The juvenile three-spined stickleback (*Gasterosteus aculeatus* L.) as a model organism for endocrine disruption II: kidney hypertrophy, vitellogenin and spiggin induction. Aquat Toxicol 70:311–326.

Hartley WR, Thiyagarajah A, Anderson MB, Broxson MW, Major SE, Zell SI (1998) Gonadal development in Japanese medaka (*Oryzias latipes*) exposed to 17β-estradiol. Mar Environ Res 46:145–148.

Herbst AL, Ulfelder H, Poskanzer DC (1971) Adenocarcinomas of the vagina: association of maternal stilbestrol therapy with tumor appearance in young women. N Engl J Med 284:878–881.

Hermesmeyer GN, Berger LL, Nash TG, Brandt Jr. RT (2000) Effects of energy intake, implantation, and subcutaneous fat end point on feedlot steer performance and carcass composition. J Anim Sci 78:825–831.

Hirai N, Nanba A, Koshio M, Kondo T, Mortia M, Tatarazako N (2006) Feminization of Japanese medaka (*Oryzias latipes*) exposed to 17-beta-estradiol: formation of testis-ova and sex-transformation during early-ontogeny. Aquat Toxicol 77:78–86.

Holbech H, Kinnberg K, Petersen GI, Jackson P, Hylland K, Norrgren L, Bjerregaard P (2006) Detection of endocrine disrupters: evaluation of a fish sexual development test (FSDT). Comp Biochem Physiol 144C:57–66.

Hook SE, Skillman AD, Small JE, Schultz IR (2006) Gene expression patterns in rainbow trout, *Oncorhyncus mykiss,* exposed to a suite of model toxicants. Aquat Toxicol 77:372–385.

Imai S, Koyama J, Fujii K (2005) Effects of 17β-estradiol on the reproduction of Java-medaka (*Oryzias javanicus*), a new test fish species. Mar Pollut Bull 51:708–714.

Imai S, Koyama J, Fujii K (2007) Effects of estrone on full life cycle of java medaka (*Oryzias javanicus*), a new marine test fish. Environ Toxicol Chem 26:726–731.

Irwin LK, Gray S, Oberdorster E (2001) Vitellogenin induction in painted turtle, *Chrysemys picta*, as a biomarker of exposure to environmental levels of estradiol. Aquat Toxicol 55:49–60.

Ivie GW, Christopher RJ, Munger CE, Coppock CE (1986) Fate and residues of [4-^{14}C]estradiol-17β after intrasmuscular injection into Holstein steer calves. J Anim Sci 62:681–690.

Jacobsen A-M, Lorenzen A, Chapman R, Topp E (2005) Persistence of testosterone and 17β-estradiol in soils receiving swine manure or municipal biosolids. J Environ Qual 34:861–871.

Jensen KM, Makynen EA, Kahl MD, Ankley GT (2006) Effects of the feedlot contaminant 17α-trenbolone on reproductive endocrinology of the fathead minnow. Environ Sci Technol 40:3112–3117.

Jones SJ, Johnson RD, Calkins CR, Dikeman ME (1991) Effects of trenbolone acetate on carcass characteristics and serum testosterone in bulls and steers on different management and implant schemes. J Anim Sci 69:1363–1369.

Kim I, Yu Z, Xiao B, Huang W (2007) Sorption of male hormones by soils and sediments. Environ Toxicol Chem 26:264–270.

Kolodziej EP, Harter T, Sedlak DL (2004) Dairy wastewater, aquaculture, and spawning fish as sources of steroid hormones in the aquatic environment. Environ Sci Technol 38:6377–6384.

Kolok AS, Snow DD, Kohno S, Sellin MK, Guillette LJ Jr (2007) Occurrence and biological effect of exogenous steroids in the Elkhorn River, Nebraska. Sci Total Environ 388:104–115.

Kramer VJ, Miles-Richardson S, Pierens SL, Giesy JP (1998) Reproductive impairment and induction of alkaline-labile phosphate, a biomarker of estrogen exposure, in fathead minnows (*Pimephales promelas*) exposed to waterborne 17β-estradiol. Aquat Toxicol 40:335–360.

Krzeminski LF, Cox BL, Gosline RE (1981) Fate of radioactive melengestrol acetate in the bovine. J Agric Food Chem 29:387–391.

Lange IG, Daxenbarger A, Schiffer B, Witters H, Ibarreta D, Meyer HHD (2002) Sex hormones originating from different livestock production systems: fate and potential disrupting activity in the environment. Anal Chem Acta 473:27–37.

Larkin P, Sabo-Attwood T, Kelso J, Denslow ND (2002) Gene expression analysis of largemouth bass exposed to estradiol, nonylphenol, and *p,p'*-DDE. Comp Biochem Physiol 133B:543–557.

Lee CY, Henricks DM, Skelley GC, Grimes LW (1990) Growth and hormonal response of intact and castrate male cattle to trenbolone acetate and estradiol. J Anim Sci 68:2682–2689.

Lee LS, Strock TJ, Sarmah AK, Rao PSC (2003) Sorption and dissipation of testosterone, estrogens, and their primary transformation products in soils and sediments. Environ Sci Technol 37:4098–4105.

LeGuevel R, Pakdel F (2001) Assessment of oestrogenic potency of chemicals used as growth promoter by in-vitro methods. Human Reprod 16:1030–1036.

Loomis AK, Thomas P (2000) Effects of estrogens and xenoestrogens on androgen production by Atlantic croaker testes *in vitro*: evidence for a nongenomic action mediated by an estrogen membrane receptor. Biol Reprod 62:995–1004.

Lone KP (1997) Natural sex steroids and the xenobiotic analogs in animal production: growth, carcass quality, pharmacokinetics, metabolism, mode of action, residues, methods and epidemiology. Crit Rev Food Sci 37:93–209.

Lorenzen A, Hendel JG, Conn KL, Bittman S, Kwabiah AB, Lazarovitz G, Masse D, McAllister TA, Topp E (2004) Survey of hormone activities in municipal biosolids and animal manures. Environ Toxicol 19:216–225.

Lorenzen A, Chapman R, Hendel JG, Topp E (2005) Persistence and pathways of testosterone dissipation in agricultural soil. J Environ Qual 34:854–860.

Matthiessen P, Arnold D, Johnson AC, Pepper TJ, Pottinger TG, Pulman KGT (2006) Contamination of headwater streams in the United Kingdom by oestrogenic hormones from livestock farms. Sci Total Environ 367:616–630.

Mellin TN, Erb RE (1966a) Estrogen metabolism and excretion during the bovine estrous cycle. Steroids 7:589–603.

Mellin TN, Erb RE (1966b) Estrogens in the bovine: a review. J Dairy Sci 48:687–700.

Meyer HHD (2001) Biochemistry and physiology of anabolic hormones used for improvement of meat production. APMIS 8:1–8.

Miles-Richardson SR, Kramer VJ, Fitzgerald SD, Render JA, Yamini B, Barbee SJ, Giesy JP (1999) Effects of waterborne exposure of 17β-estradiol on secondary sex characteristics and gonads of fathead minnows (*Pimephales promelas*). Aquat Toxicol 47:129–145.

Miller DH, Jensen KM, Villeneuve DL, Kahl MD, Makynen EA, Durhan EJ, Ankley GT (2007) Linkage of biochemical responses to population-level effects: a case study with vitellogenin in the fathead minnow (*Pimephales promelas*). Environ Toxicol Chem 26:521–527.

Mills LJ, Gutjahr-Gobell RE, Horowitz DB, Denslow ND, Chow MC, Zaroogian GE (2003) Relationship between reproductive success and male plasma vitellogenin concentrations in cunner, *Tautogolabrus adspersus*. Environ Health Perspect 111:93–99.

Miracle A, Ankley G, Lattier D (2006) Expression of two vitellogenin genes (vg1 and vg3) in fathead minnow (*Pimephales promelas*) liver in response to exposure to steroidal estrogens and androgens. Ecotoxicol Environ Saf 63:337–342.

Miura T, Higuchi M, Ozaki Y, Ohta T, Miura C (2006) Progestin is an essential factor for the initiation of the meiosis in spermatogenetic cells of the eel. Proc Natl Acad Sci U S A 103:7333–7338.

Moens LN, van der Ven K, Van Remortel P, Del-Favero J, De Coen WM (2006) Expression profiling of endocrine-disrupting compounds using a customized *Cyprinus carpio* cDNA microarray. Toxicol Sci 93:298–310.

Montgomery TH, Dew PF, Brown MS (2001) Optimizing carcass value and the use of anabolic implants in beef cattle. J Anim Sci 79(suppl E):E296–E306.

Nayak PK, Mishra J, Ayyappan S, Singh BN (2001) 17α-Hydroxyl progesterone-induced breeding of the stinging catfish *Heteropneustes folssilis* (Bolch) with or without priming of gonadotropin. J Aquat Trop 16:159–164.

Nichols DJ, Daniel TC, Moore Jr. PA, Edwards DR, Pote DH (1997) Runoff of estrogen hormone 17β-estradiol from poultry litter applied to pasture. J Environ Qual 26:1002–1006.

Orlando EF, Kolok AS, Binzcik G, Gates J, Horton MK, Lambright C, Gray LE, Guillette LJ (2004) Endocrine disrupting effects of cattle feedlot effluent on an aquatic sentinel species, the fathead minnow. Environ Health Perspect 112:353–358.

Orn S, Yamani S, Norrgren L (2006) Comparison of vitellogenin induction, sex ratio, and gonad morphology between zebrafish and Japanese medaka after exposure to 17α-ethinylestradiol and 17β-trenbolone. Arch Environ Contam Toxicol 51:237–243.

Panter GH, Thompson RS, Sumpter JP (1998) Adverse reproductive effects in male fathead minnows (*Pimephales promelas*) exposed to environmentally relevant concentrations of the natural oestrogens, oestradiol and oestrone. Aquat Toxicol 42:243–253.

Parliamentary Office of Science and Technology, House of Commons, London, UK (1999) Hormones in beef. <http://www.parliament.uk/post/pn127.pdf>.

Pelissero C, Flouriot G, Foucher JL, Bennetau B, Dunogues J, LeGac F, Sumpter JP (1993) Vitellogenin synthesis in cultured hepatocytes; an *in vitro* test for the estrogenic potency of chemicals. J Steroid Biochem Mol Biol 44:263–272.

Peterson EW, Davis RK, Orndorff HA (2000) 17β-Estradiol as an indicator of animal waste contamination in mantled karst aquifers. J Environ Qual 29:826–834.

Pottier J, Cousty C, Heitzman RJ, Reynolds P (1981) Differences in the biotransformation of a 17b-hydroxylated steroid, trenbolone acetate in rat and cow. Xenobiotica 11:489–500.

Preston RL (1997) Rationale for the safety of implants. <http://www.fortdodgelivestock. com/pdfs/Pdfs%20for%20beef/RationaleforSafety(241).pdf>.

Purdy RH, Durocher CK, Moore Jr. PH, Rao PN (1980) Analysis of metabolites of progesterone in bovine liver, kidney, kidney fat, and milk by high performance liquid chromatography. J Steroid Biochem 12:1307–1315.

Raloff J (2002) Hormones: here's the beef. Science News 161:10–12.

Renner R (2002) Do cattle growth hormones pose an environmental risk? Environ Sci Technol 36:194A–197A.

Rico AG (1983) Metabolism of endogenous and exogenous anabolic agents in cattle. J Anim Sci 57:226–232.

Rumsey TS, Tyrrell HF, Dinius DA, Moe PW, Cross HR (1981) Effect of diethylstilbestrol on tissue gain and carcass merit of feedlot beef steers. J Anim Sci 53:589–600.

Schiffer B, Daxenberger A, Meyer K, Meyer HHD (2001) The fate of trenbolone acetate and melengestrol acetate after application as growth promoters in cattle: environmental Studies. Environ Health Perspect 109:1145–1151.

Schwarzenberger F, Mosel E, Pamel R, Bamberg E (1996) Faecal steroid analysis for non-invasive monitoring of reproductive status in farm, wild and zoo animals. Anim Reprod Sci 42:515–526.

Seki M, Fujishima S, Nozaka T, Maeda M, Kobayashi K (2006) Comparison of response to 17β-trenbolone among three small fish species. Environ Toxicol Chem 25:2742–2752.

Shioda T, Wakabayashi M (2000) Effect of certain chemicals on the reproduction of medaka (*Oryzias latipes*). Chemosphere 40:239–243.

Shore LS, Gurevitz M, Shemesh M (1993) Estrogen as an environmental pollutant. Bull Environ Contam Toxicol 51:361–366.

Siemens MG (1996) Tools for optimizing feedlot production. Publication A366. Cooperative Extension, University of Wisconsin, Madison, WI.

Sone Z, Hinage M, Itamoto M, Katsu Y, Watanabe H, Urushitani H, Tooi O, Guillette LJ Jr, Iguchi T (2005) Effects of an androgenic growth promoter 17β-trenbolone on masculinization of mosquitofish (*Gambusia affinis affinis*). Gen Comp Endocrinol 143:151–160.

Soto AM, Calabro JM, Prechtl NV, Yau AY, Orlando EF, Daxenberger A, Kolok AS, Guillette LJ Jr, le Bizec B, Lange IG, Sonnenschein C (2004) Androgenic and estrogenic activity in cattle feeldot effluent receiving water bodies of eastern Nebraska, USA. Environ Health Perspect 112:346–352.

Sundararaj BI, Panchanan N (1981) Steroid-induced synthesis of vitellogenin in the catfish, *Heteropneustes folssilis* (Bolch). Gen Comp Endocrinol 43:201–210.

Thorpe KL, Cummings RI, Hutchinson TH, Scholze M, Brighty G, Sumpter JP, Tyler CR (2003) Relative potencies and combination effects of steroidal estrogens in fish. Environ Sci Technol 37:1142–1149.

Wade N (1972) DES: A case of regulatory abdication. Science 177:335–337.

ZoBell D, Chapman CK, Heaton K, Birkelo C (2000) Beef cattle implants. Utah State University Extension Electronic Publication AG-509. <http://extension.usu.edu/htm/ publications/publication=5931>.

Arsenic in Marine Mammals, Seabirds, and Sea Turtles

Takashi Kunito, Reiji Kubota, Junko Fujihara, Tetsuro Agusa, and Shinsuke Tanabe

T. Kunito, R. Kubota, J. Fujihara, T. Agusa, S. Tanabe(✉)
Center for Marine Environmental Studies (CMES), Ehime University, Bunkyo-cho 2-5, Matsuyama 790-8577, Japan (shinsuke@agr.ehime-u.ac.jp)

T. Kunito
Department of Environmental Sciences, Faculty of Science, Shinshu University, 3-1-1 Asahi, Matsumoto 390-8621, Japan.

R. Kubota
Division of Environmental Chemistry, National Institute of Health Sciences, Kamiyoga 1-18-1, Setagaya-ku, Tokyo 158-8501, Japan.

J. Fujihara
Department of Legal Medicine, Shimane University School of Medicine, 89-1 Enya, Izumo, Shimane 693-8501, Japan.

D.M. Whitacre (ed.), *Reviews of Environmental Contamination and Toxicology.*

1 Introduction

Arsenic, a chalcophilic element, is widespread in the environment. Although arsenic may possibly be an essential element for life (Cox 1995) and some microorganisms are known to use arsenic for energy generation (Oremland and Stolz 2003), no firm data are available on its essentiality for biological systems (Francesconi 2005). In contrast to its possible essentiality in life, many studies have focused on its high toxicity, which has been well known from various cases of poisoning throughout the ages (Nriagu 2002). The toxicity is especially high for inorganic arsenic; trivalent inorganic arsenic [arsenite; As(III)] is known to bind readily to sulfhydryl groups of enzymes leading to enzyme inhibition, whereas pentavalent inorganic arsenic [arsenate; As(V)], which is structurally similar to phosphate, may disrupt metabolic reactions that require phosphorylation (Cox 1995). Symptoms of acute intoxication in humans by inorganic arsenic include severe gastrointestinal disorders, hepatic and renal failure, and cardiovascular disturbances, whereas chronic exposure causes skin pigmentation, hyperkeratosis, and cancers in the lung, bladder, liver, and kidney as well as skin (Gorby 1994; WHO 2001). At present, arsenic contamination of groundwater is a worldwide problem (Mandal and Suzuki 2002), particularly in the Bengal Delta where chronic ingestion of arsenic in groundwater poses a significant health risk to about 36 million people (Nordstrom 2002). Thus, the development and use of techniques to remove arsenic from polluted groundwater is an urgent necessity (Chowdhury 2004). In contrast to the hazards of arsenic, it is useful in medicine. For example, arsenic trioxide (As_2O_3) has recently attracted considerable attention as a therapeutic agent for treatment of acute promyelocytic leukemia and other cancers, although the precise mechanisms by which it produces results are not fully understood (Zhu et al. 2002).

Arsenic is used in agriculture, livestock, medicine, electronics, industry, and metallurgy (Azcue and Nriagu 1994). Worldwide anthropogenic emission of arsenic was estimated to be ~5,000 t/yr in the mid-1990s, of which more than half was accounted for by nonferrous metal production (Pacyna and Pacyna 2001). Emission from natural sources, estimated to be ~12,000 t/yr (Pacyna and Pacyna 2001), is more than twice that from anthropogenic sources. The major natural source is volcanoes (Nriagu 1989). Therefore, both anthropogenic and natural sources should be factored into evaluations when assessing the environmental risk of arsenic.

Generally, no significant difference is observed between arsenic concentrations in seawater and freshwater: arsenic concentration is about 1.5 μg L^{-1} in seawater, 0.1–2.0 μg L^{-1} in river water with absence of significant nearby emission sources (e.g., mining activity and geothermal sources) and < 1 μg L^{-1} in lake water (Plant et al. 2005). However, it is widely known that marine organisms contain arsenic at much higher concentrations than do terrestrial organisms (Lunde 1977), and some marine species show arsenic levels exceeding 2,000 μg g^{-1} dry wt on a whole-body basis (Gibbs et al. 1983). Hence, many studies have been conducted on arsenic levels and its speciation in marine organisms at low trophic levels (e.g., algae and shellfish). In contrast, few studies are available for marine mammals and seabirds occupying higher trophic levels, or sea turtles. It is known that marine mammals and seabirds accumulate organochlorine compounds (e.g., polychlorinated biphenyls) at high levels by biomagnification through the marine food chain (O'Shea 1999; Braune et al. 2005; Tanabe and Subramanian 2006). In particular, marine mammals have a unique tissue, blubber, which serves as the main repository for organochlorine compounds. Furthermore, it has been reported that marine mammals (Thompson 1990; Law 1996; O'Shea 1999), seabirds (Thompson 1990), and sea turtles (Anan et al. 2001) accumulate certain metals, such as cadmium, mercury, and copper, at high concentrations in their tissues. For example, some marine mammals show hepatic mercury levels of 13,000 μg g^{-1} dry wt and a renal cadmium level of 800 μg g^{-1} dry wt (O'Shea 1999). Such high accumulation of metals seems to depend not only on biomagnification through the food chain but also on various biological factors, such as species, feeding habits, and lifespan (Thompson 1990). Indeed, it has been suggested that organic mercury is virtually the only metal that can be biomagnified through the food chain (Langston and Spence 1995). Although many studies have been conducted on the accumulation of metals such as cadmium, copper, mercury, and zinc in tissues of marine mammals, seabirds, and sea turtles, there have been few efforts to study the presence of arsenic species in these animals. To illustrate, although more than 18,000 papers have been published on the accumulation of organochlorine compounds and metals in marine mammals since the 1960s (O'Shea and Tanabe 2003), no report was published on arsenic species in marine mammals (Eisler 1994; Law 1996) until the study of Goessler et al. (1998); this is probably because sensitive speciation techniques for arsenic were unavailable for many years. Because marine mammals, seabirds, and sea turtles display unique features in metal accumulation, it may be useful to characterize arsenic accumulation in these animals. Furthermore, studying arsenic species and their presence in high-trophic-level marine animals is crucial for understanding arsenic cycling in the marine ecosystem. In this review, we focus attention on the pattern of accumulation of arsenic species in marine mammals, seabirds, and sea turtles and also summarize the state of current knowledge related to this topic (e.g., newly identified arsenicals in other marine organisms).

2 Arsenic Species and Cycling in the Marine Ecosystem

2.1 Arsenic Species

Arsenic is present in various chemical forms (Fig. 1), and its toxicity depends on the particular chemical form. Therefore, an understanding of arsenic speciation is essential to understanding its environmental behavior and ecotoxicological effects. Recently, the new scientific field, "metallomics," which focuses on identification of metallomes (metalloproteins, metalloenzymes, and other metal-containing biomolecules)

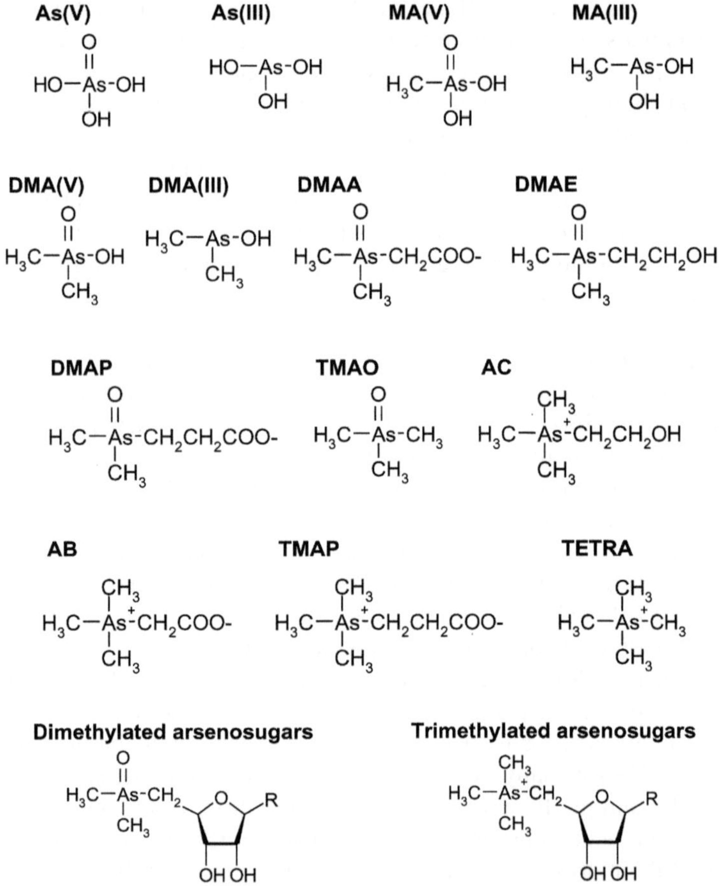

Fig. 1 Water-soluble arsenicals found in the marine ecosystem: As(V), arsenate; As(III), arsenite; MA(V), methylarsonic acid; MA(III), methylarsonous acid; DMA(V), dimethylarsinic acid; DMA(III), dimethylarsinous acid; DMAA, dimethylarsinoyl acetate; DMAE, dimethylarsinoyl ethanol; DMAP, dimethylarsinoyl propionate; TMAO, trimethylarsine oxide; AC, arsenocholine; AB, arsenobetaine; TMAP, trimethylarsoniopropionate; TETRA, tetramethylarsonium ion

and the elucidation of their functions in biological systems, has received increasing attention (Haraguchi 2004). Arsenic speciation is also of interest from the point of view of this new field (Suzuki 2005). High performance liquid chromatography-inductively coupled plasma-mass spectrometry (HPLC-ICP-MS) is the technique most employed to determine arsenic speciation. Arsenic speciation is conducted using ICP-MS as a detector after separation of arsenic species by HPLC. The HPLC-ICP-MS technique is the most sensitive method for detecting arsenic species and often allows a detection limit lower than 0.5 $\mu g\ L^{-1}$ (e.g., Hirata et al. 2006). The choice of the HPLC column (e.g., anion-exchange, cation-exchange, or reversed-phase) affects the separation of arsenic species, because different arsenic species behave differently on each column. In part, this variability may occur because the dissociation constant, functional groups, and molecular size are largely different among arsenic species. Therefore, HPLC columns should be selected based on characteristics of the expected analytes and coexisting arsenicals. It should be noted that the HPLC-ICP-MS technique requires the appropriate authentic reference material for successful identification of arsenic species because no structural information is inherently provided by this method. Identification of arsenic species depends on a comparison of the retention times of unknowns versus such authentic standards using selected HPLC column(s).

Seawater is considered as the starting point for arsenic cycling in the marine ecosystem. In seawater, most arsenic exists in the inorganic form (Fig. 2), with pentavalent arsenic, $HAsO_4^{2-}$, predominating in oxygenated surface water (Cullen and Reimer 1989). Arsenic shows a nutrient-like vertical profile in the water column (i.e., depletion at the euphotic zone), suggesting biological uptake of arsenic

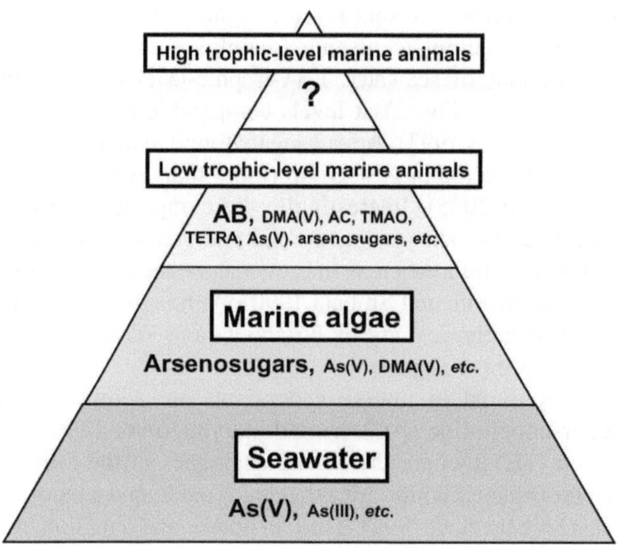

Fig. 2 Arsenicals found in seawater and organisms at each trophic level

by marine phytoplankton, in spite of its high toxicity (Cullen and Reimer 1989; Shibata and Morita 2000). In addition to inorganic arsenic, pentavalent monomethylated and dimethylated arsenicals, methylarsonic acid [MA(V)] and dimethylarsinic acid [DMA(V)], respectively, are also present. Santosa et al. (1996) reported that the ratio of MA(V) + DMA(V) to total arsenic increased with water temperature and also was influenced by nutrient levels in Pacific Ocean surface waters, which suggests that the abundance of organoarsenicals reflect the biological activity [i.e., uptake of As(V) and subsequent methylation to MA(V) and DMA(V)] of phytoplankton in the surface water. Furthermore, trivalent methylarsonous acid [MA(III)] and dimethylarsinous acid [DMA(III)] were also detected in seawater (Hasegawa 1996). It should be noted that the methylation pathway of inorganic arsenic is not yet firmly established. It has been generally accepted that inorganic arsenic is methylated oxidatively (Fig. 3a), but recently a reductive methylation pathway has also been proposed (Hayakawa et al. 2005; Naranmandura et al. 2006). In the latter pathway, MA(V) and DMA(V) are shown as the end products of transformation (Fig. 3b), which is consistent with the abundant presence of these pentavalent arsenicals in animals (Aposhian and Aposhian 2006).

In marine organisms, arsenic is known to exist mainly as organic forms, although elucidation of the actual structures involved only took place over many years. In 1977, arsenobetaine (AB; see Fig. 1) was first identified in the western rock lobster (*Panulirus cygnus*) (Edmonds et al. 1977). AB was first synthesized in the 1930s for pharmacological studies, but its presence in biota and the environment was not reported until 1977 (Edmonds et al. 1993). In 1981, arsenosugars (Fig. 1) were also identified in brown kelp, *Ecklonia radiata* (Edmonds and Francesconi 1981). Subsequent studies on arsenic species in various low-trophic-level marine organisms revealed that marine algae, which rest at the base of the marine food chain, accumulate arsenic (mainly as arsenosugars) at levels of 1,000–50,000 times that of seawater. Low-trophic-level marine animals contain arsenic mainly as AB (see Fig. 2) at levels comparable to those in marine algae (Francesconi and Edmonds 1993). Arsenosugars found in marine algae and in some marine animals comprise the largest group (more than 20) of naturally occurring arsenicals (Francesconi 2005). Interestingly, the composition of arsenosugars in marine algae is related to their phylogeny: red and green algae contain arsenosugars of rather simple structure, whereas in brown algae the structure of arsenosugars is more complicated (Morita and Shibata 1990). Although AB was not detected in marine algae until recently, a study by Nischwitz and Pergantis (2005a) revealed the presence of AB in these organisms.

Major arsenicals found in marine ecosystems are shown in Figs. 1 and 2. Arsenobetaine, arsenocholine (AC), trimethylarsine oxide (TMAO), and tetramethylarsonium ion (TETRA) are the arseno-analogues of the nitrogen-containing compounds, glycine betaine, choline, trimethylamine oxide, and tetramethylammonium ion, respectively (Shibata et al. 1992). Thus, in uptake and retention, marine animals do not discriminate these arsenicals from their natural nitrogen analogues (Shibata et al. 1992).

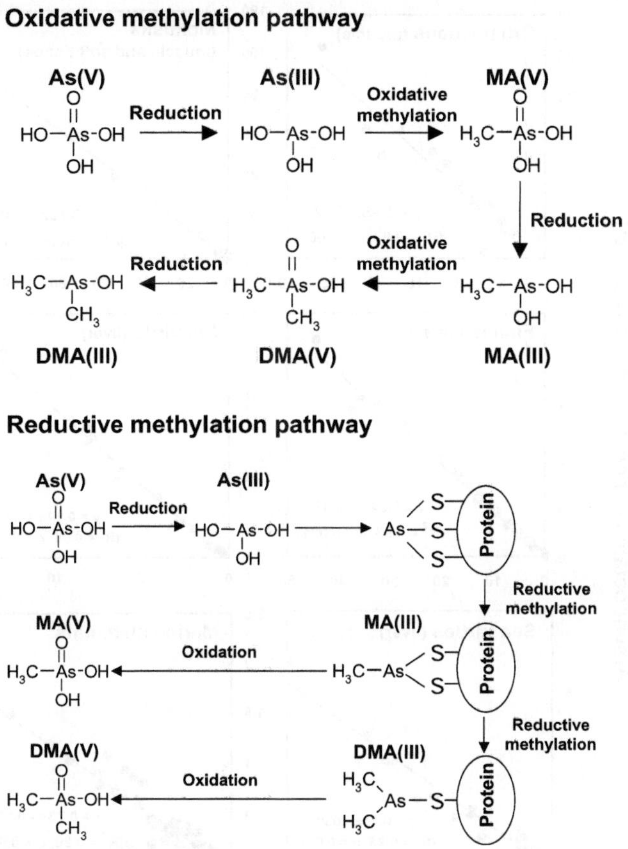

Fig. 3 Hypothesized oxidative (a) and reductive (b) methylation pathways of inorganic arsenic

Marine animals generally contain arsenic mainly as AB (Figs. 2, 4), with two exceptions: the first, a marine teleost fish, the silver drummer (*Kyphosus sydneyanus*), which digests its macroalgal diet by fermentation (Edmonds et al. 1997), and second, the dugong (*Dugong dugon*), which feeds on seagrass (Kubota et al. 2002a, 2003b). The proportion of AB in marine animals varies depending on their feeding habit and trophic position, with animals of higher trophic levels containing higher proportions of AB (Francesconi and Kuehnelt 2002). For example, AB comprises the major arsenic species in pelagic carnivorous marine fish, whereas various arsenicals are contained in detritivorous and herbivorous marine fish, with the corresponding proportion of AB being relatively low (Kirby and Maher 2002). In general, the arsenic composition in marine animals reflects the distribution found in their prey, because marine animals take up arsenicals mainly through their diets (Phillips 1990). However, the trophic transfer coefficient differs among species of arsenicals. Although inorganic arsenic predominates in seawater, dimethylated arsenosugars and

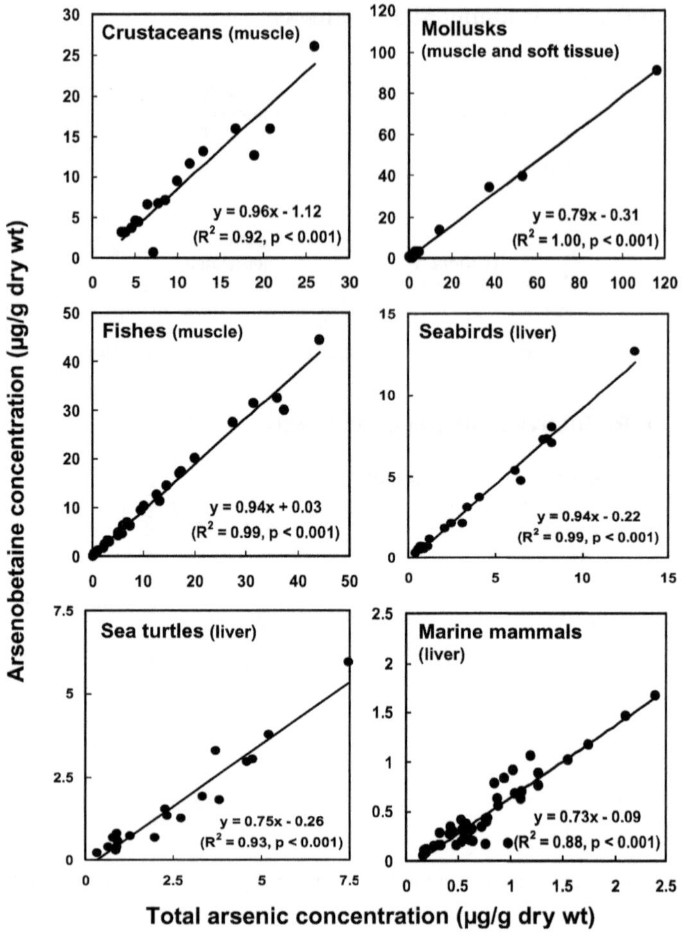

Fig. 4 Relationship between total arsenic and arsenobetaine concentrations in marine animals. Total arsenic in marine mammals, seabirds, and sea turtles includes nonextractable arsenic in liver. Data on crustaceans, mollusks, and fishes are from references cited in Francesconi and Edmonds (1993); hepatic arsenic concentrations of marine mammals are from Kubota et al. (2003a), Fujihara et al. (2003), and Goessler et al. (1998); those of seabirds are from Kubota et al. (2003a) and Fujihara et al. (2003); and those of sea turtles are from Kubota et al. (2003a) and Fujihara et al. (2003)

trimethylated arsenicals (i.e., AB) are contained as major arsenicals in marine algae and marine animals, respectively (Fig. 2), with proportions of more methylated forms increasing with trophic level. Hence, the proportion of AB and degree of methylation of predominant arsenicals are related to the trophic position of the organism. However, the concentrations of total arsenic and AB vary greatly among species (Francesconi and Edmonds 1993) and are not related to the trophic position of the organism (Table 1).

Table 1 Arsenic concentrations in marine organisms (μg g^{-1} dry wt)

Organism	Range of means (Phillips 1990)	Geometric means (Neff 1997)	Means (Kubota et al. 2001)	Range of means (Saeki et al. 2000)	Range of means (Kubota et al. 2003b)
Marine algae	1.5–84	43.7			
Crustaceans	8.4–179	14.9			
Bivalve mollusks	5.0–1025	10.4			
Gastropod mollusks	9.0–407	52.0			
Cephalopod mollusks	6.4–99	16.1			
Fish	<0.2–216	5.6			
Marine mammals					
Pinnipeds (liver)			1.85		
Cetaceans (liver)			1.88		
Sea turtles (liver)				1.76–15.3	
Seabirds (liver)					2.25–12.2

The widely used HPLC-ICP-MS technique is effective in identification and quantification of arsenicals for which corresponding authentic standards are available, but it is not applicable in the absence of such standards. Recently, electrospray tandem mass spectrometry (ES-MS/MS) has been increasingly used for identification of arsenicals without using standard compounds, and has, therefore, contributed significantly to the understanding of arsenic metabolism in marine organisms and arsenic cycling in marine ecosystems (Edmonds and Francesconi 2003). For example, McSheehy et al. (2002) identified 15 organoarsenicals in the kidney of giant clams (*Tridacna derasa*), which have symbiotic unicellular algae in their tissues. These identifications were achieved using ES-MS/MS after successive chromatographic fractionation of the arsenicals by size-exclusion chromatography, anion-exchange chromatography, and cation-exchange chromatography (or anion-exchange chromatography with a high resolution column). More recently, Nischwitz and Pergantis (2006) established a HPLC-ES-MS/MS method that is capable of analyzing for 50 arsenic species, including various thio-arsenicals. Furthermore, in a recent study using ES-MS/MS, dimethylarsinoyl acetate (DMAA), dimethylarsinoyl propionate (DMAP) and dimethylarsinoyl ethanol (DMAE), which are postulated intermediates in AB biosynthesis, were found in various marine animals and marine algae (Sloth et al. 2005a). Also, thio-arsenosugars containing As=S have been recently identified in mussels and marine algae using this method (Fricke et al. 2004; Schmeisser et al. 2004; Nischwitz et al. 2006) and HPLC-ICP-MS (Meier et al. 2005). It is noteworthy that these newly identified arsenicals have also been quantified for certified reference materials (CRMs) from marine animals (tuna fish, BCR-627; dogfish muscle, DORM-2; mussel tissue, CRM278R and oyster, 1566b) (Nischwitz and Pergantis 2005b)

2.2 Microbial Degradation of Arsenobetaine

Because AB is by far the dominant arsenical in most marine animals, degradation of AB to inorganic arsenic after its release into the environment from the decomposing dead animals is essential for completion of arsenic cycling in marine ecosystems. There are two possible pathways for AB degradation (Fig. 5): the conversion from AB to TMAO or from AB to DMAA. The TMAO or DMAA is further degraded to inorganic arsenic through DMA(V) in both pathways. The AB-degrading bacteria are ubiquitous in the marine environment. It was shown that microbial communities from marine sediments, marine algae, mollusk intestine, or suspended particles were able to convert AB to TMAO, DMA(V), and even to inorganic arsenic (Hanaoka et al. 1992). Microbial communities on suspended particles collected at a depth of 3500 m were also able to degrade AB (Hanaoka et al. 1997). It is likely that aerobic microorganisms are primarily involved in degradation of AB, because AB is more rapidly degraded under aerobic than anaerobic conditions (Hanaoka et al. 1992). Khokiattiwong et al. (2001) suggested that AB is rapidly degraded when present at its environmentally relevant low level. In contrast, degradation took several weeks in an incubation experiment conducted by Hanaoka et al. (1992) in which a relatively high level of AB was employed. More than 95% of AB was converted to DMA(V) within 24 hr by microorganisms in seawater to which low levels of AB (100 and 750 μg As L^{-1}) were added and in which shore crabs (*Carcinus maenas*) were maintained (Khokiattiwong et al. 2001). A more detailed investigation revealed that AB was first converted to DMAA, reaching a maximum concentration after 3 hr incubation, and was then totally converted to DMA(V) after 48 hr. Thus, the authors expect that AB is not usually detected in seawater because of such rapid degradation. In these experiments, TMAO was not detected, suggesting that AB was degraded to DMA(V) primarily via DMAA.

In addition to studies on AB degradation by microbial communities, isolation and characterization of each AB-degrading bacterium have also been reported. Two

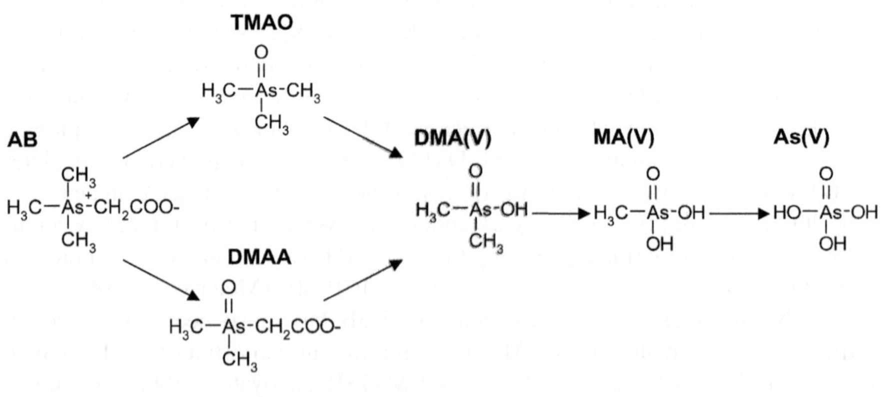

Fig. 5 Hypothesized degradation pathways of arsenobetaine

bacterial strains of the *Vibrio-Aeromonas* group isolated from coastal sediment by the culture enrichment method converted AB to DMA(V) under aerobic but not anaerobic conditions (Hanaoka et al. 1992). A microbial community isolated from the blue mussel (*Mytilus edulis*) converted AB to TMAO, MA(V), and DMA(V). Four AB-degrading bacterial strains (one of *Paenibacillus*, two of *Pseudomonas*, and one of *Aeromonas*) were isolated from this community to characterize the degradation pathway of AB (Jenkins et al. 2003). Degradation of AB to DMA(V) by each strain occurred after 21 d incubation. TMAO was detected during incubation with the microbial community, whereas DMAA but not TMAO was observed during incubation in the pure culture. One isolate further degraded DMAA to As(V) after 28 d incubation. Jenkins et al. (2003) assumed that the conversion of AB to DMAA would be a fortuitous reaction, whereas conversion of DMAA to DMA(V) would provide carbon or energy to the bacteria. Also, the degradation of AB was shown to be mediated intracellularly by *Paenibacillus* sp. (Jenkins et al. 2003). Devesa et al. (2005) observed that microbial communities from the hepatopancreas, tail, and remaining parts of the red swamp crayfish (*Procambarus clarkii*) degraded AB to TMAO, DMA(V), MA(V), and an unidentified arsenical. Interestingly, in the incubation experiments using either AC, TETRA, TMAO, DMA(V), or MA(V), only AC was converted to AB, but the other arsenicals were not transformed by these microbial communities. Five AB-degrading strains isolated from these microbial communities were all identified as *Pseudomonas putida* and were shown to degrade AB to DMA(V) and MA(V) in the incubation experiment (Devesa et al. 2005). Generally, microbial communities could degrade AB to inorganic arsenic, whereas most of the AB-degrading bacteria could not degrade AB completely by themselves (Hanaoka et al. 1992). Thus, degradation of AB to inorganic arsenic requires the cooperation of various microorganisms. The microbial community structure in AB degradation is important, because metabolites formed by the degradation are different among the sources of microbial communities (Hanaoka and Kaise 1999).

3 Distribution of Arsenic Species in the Tissues of Marine Mammals, Seabirds, and Sea Turtles

3.1 Arsenic in Marine Mammals

There are few studies that have examined the types of arsenic species in marine mammals, seabirds and sea turtles. Therefore, we have undertaken a detailed characterization of arsenic accumulation in such large marine animals.

Influences of feeding habits, age (or body size), and gender on the hepatic arsenic level in marine mammals were examined by analyzing in-house measurements of 16 species of marine mammals ($n = 226$), as well as data from the literature (Kubota et al. 2001). The highest level of 7.68 μg g^{-1} on a dry weight (dry wt) basis was observed in liver of the harp seal (*Pagophilus groenlandicus*); levels were lower than those for animals at lower trophic levels (see Table 1). Hepatic levels were comparable between pinnipeds and cetaceans (Table 1). Influences of gender and age (or body size) on the arsenic level were not found (Kubota et al. 2001). The relatively low arsenic level in marine mammals is probably because arsenic is mainly present as AB, which has a short biological half-life in marine mammal tissues. However, hepatic levels in marine mammals vary by species and depend on feeding habits (Fig. 6); species feeding on cephalopods and crustaceans tend to contain higher arsenic concentrations than those feeding on fish, which is consistent with the pattern observed in prey organisms (Table 1). Generally, concentrations of other trace elements also vary with feeding habits. For example, marine mammals feeding on cephalopods show higher concentrations of cadmium and radioactive cesium, whereas animals feeding on fish exhibit higher mercury levels (Watanabe et al. 2002; Yoshitome et al. 2003).

Goessler et al. (1998) were among the first to report arsenic species in marine mammals. These authors found AB to be the predominant arsenical in all the liver samples of the ringed seal (*Pusa hispida*; $n = 10$), bearded seal (*Erignathus barbatus*; $n = 1$), pilot whale (*Globicephala melas*; $n = 2$), and beluga (*Delphinapterus leucas*; $n = 1$), accounting for 68%–98% of extractable arsenic. Arsenocholine and DMA(V) were also found in almost all samples, whereas MA(V) was detected only in 5 specimens. Because TETRA was detected in pinnipeds but not in cetaceans,

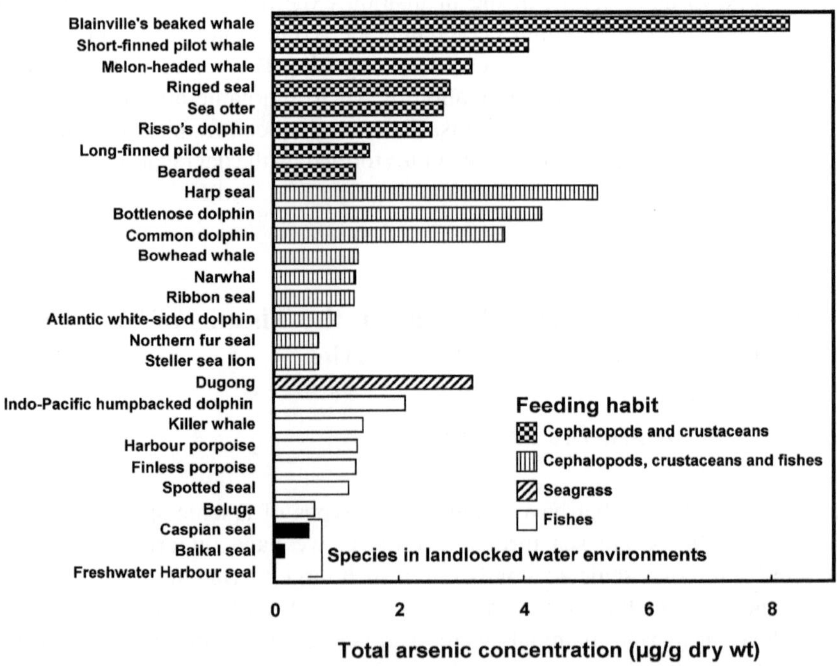

Fig. 6 Influence of feeding habits on arsenic concentration in liver of marine mammals (Kubota et al. 2001)

these authors hypothesized that the presence or absence of TETRA reflects differences in metabolism between these two groups of marine mammals (Goessler et al. 1998). However, the limited number of samples used in their study (only 3 for cetaceans) made it difficult to draw a definitive conclusion about the origin of TETRA. Arsenic speciation analyses were performed in liver of Dall's porpoise (*Phocoenoides dalli*), short-finned pilot whale (*Globicephala macrorhynchusfive*), harp seal, ringed seal, and dugong, also by Kubota et al. (2002a, 2003a). Arsenobetaine was the dominant arsenical in all samples tested except in the dugong. Lower AB (42%) and higher DMA(V) percentages (38%) were found in the Dall's porpoise than in other species (Fig. 7). Although TETRA was not detected in cetaceans by Goessler et al. (1998), this arsenical was found in the Dall's porpoise and also in pinnipeds, the harp seal, and the ringed seal (Kubota et al. 2003a), indicating that TETRA is present in both cetaceans and pinnipeds. Interestingly, AB was present in only a trace amount in the dugong ($n = 1$), with MA(V) being the major arsenical followed by DMA(V) (Kubota et al. 2003a). To confirm the generality of this unique composition of arsenicals in the dugong, arsenic speciation was conducted in four additional liver samples of the dugong (Kubota et al. 2003b); the animals did not contain AB at a detectable level and had appreciable amounts of MA(V) and smaller quantities of DMA(V) (Kubota et al. 2003b), which was in agreement with the result of Kubota et al. (2003a). These results can be attributed to the seagrass diet of the dugong. Although only limited data are available, it is believed that, in contrast to marine algae, seagrass might not contain arsenosugars (Shibata and Morita 2000). Because seagrass is phylogenetically related to terrestrial higher plants rather than to marine algae, it may contain principally MA(V) and DMA(V) as do terrestrial plants (Kuehnelt and Goessler 2003).

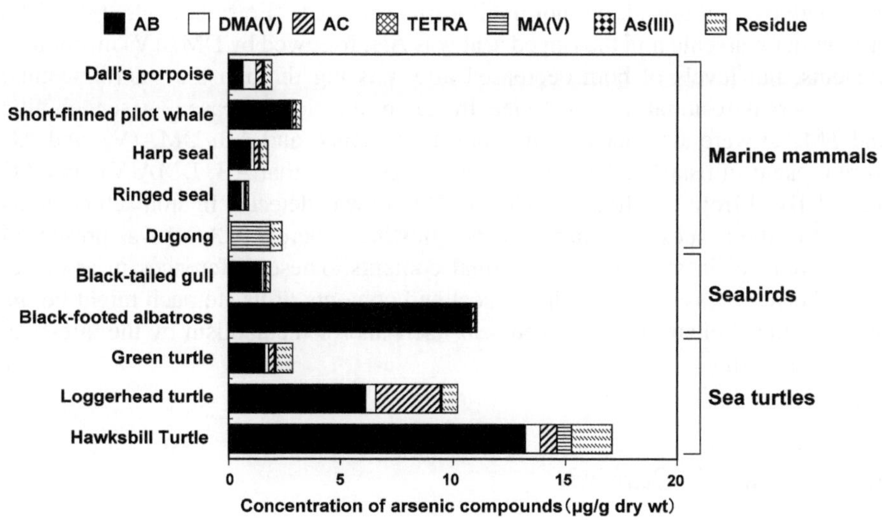

Fig. 7 Arsenic species in liver of marine mammals, seabirds, and sea turtles (Kubota et al. 2003a; Fujihara et al. 2003)

Inorganic arsenic was not detected in the liver samples of marine mammals we examined. Sloth et al. (2005b) proposed a new method for determining inorganic arsenic in animal samples that probably successfully extracts As(III) bound to thiol groups of proteins. In their procedure, all inorganic arsenic was determined as As(V) by HPLC-ICP-MS after microwave-assisted alkaline digestion of the sample [oxidizing As(III) to As(V) in alkaline media]. Inorganic arsenic was not detected in the minke whale (*Balaenoptera acutorostrata*), harp seal, and hooded seal (*Cystophora cristata*), even though the aforementioned procedure was used (the tissue was not mentioned in the paper, but it was probably muscle; Sloth et al. 2005b).

Almost all studies on arsenic have focused on its speciation in the liver (the main metabolic organ) of marine mammals, but little is known about the distribution of arsenic species in other tissues. Ebisuda et al. (2002) analyzed arsenic species in liver, kidney, muscle, and gonad and total arsenic in blubber and hair of the ringed seal ($n = 18$), and found that arsenic levels were highest in blubber, followed by liver and kidney, and lowest in muscle, gonad, and hair on a wet weight basis. Assuming that the respective tissue weight ratio of liver:kidney:muscle:blubber: hair is 10:1:100:200:5, about 90% of the arsenic burden of the five tissues is estimated to be present in ringed seal blubber. It is reported that the forms of arsenic differ between dogfish muscle and liver (Wahlen et al. 2004). In these samples, AB accounted for 96% of arsenic in muscle, and AB and DMA(V), respectively, accounted for 79% and 16% of arsenic in liver, suggesting differences in arsenic metabolism between the two tissues. However, there was no such difference in the ringed seal, where AB accounted for more than 70% of extractable arsenic in all the liver, kidney, muscle, and gonad (Ebisuda et al. 2002). It should be noted that lipid-soluble arsenicals prevail only in the blubber, accounting for about 90% of total arsenic (Ebisuda et al. 2002). Interestingly, AC was detected in all samples of liver, kidney, and gonad, but not in all muscle samples (Ebisuda et al. 2002). The predominant arsenical in the ringed seal was AB, followed by DMA(V) in stomach contents, but levels of both decreased after passing through the gastrointestinal tract whereas residual arsenic (nonextractable arsenic) increased. Arsenocholine and TMAO were also detected in stomach contents, and AB, DMA(V), and AC were present in tissues of the ringed seal, suggesting that AB, DMA(V), and AC were derived from the diet. In contrast, TMAO was detected in stomach contents but not in the tissues or contents of the intestine, whereas MA(V) was present in tissues but not in stomach or intestinal contents. These differences in arsenical forms between tissues of the ringed seal and contents of its stomach might be the result of metabolism by the ringed seal itself and/or metabolism by the intestinal bacteria it harbors.

3.2 Arsenic in Seabirds

Arsenic levels are higher in seabirds than terrestrial birds (Fig. 8), although there are few studies that define arsenic species in birds. Arsenic speciation in birds was first reported

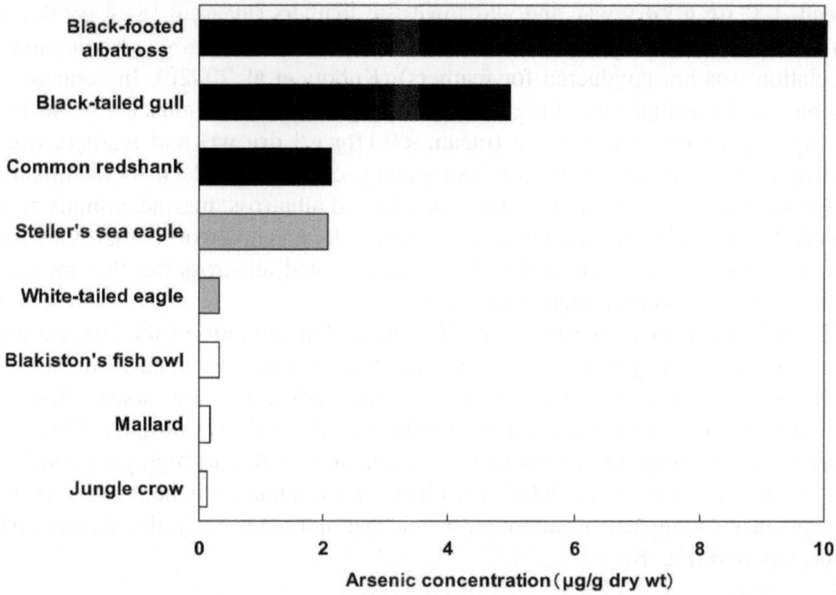

Fig. 8 Comparison of hepatic arsenic concentrations among birds from marine, coastal, and terrestrial environments (Kubota et al., unpublished results). *Filled*, *shaded*, and *open bars* correspond to birds from marine, coastal, and terrestrial environments, respectively

by Kubota et al. (2002b, 2003a). These authors reported average arsenic levels up to 12.2 μg g^{-1} dry wt, and up to 26.7 μg g^{-1} dry wt in liver of the black-footed albatross (*Phoebastria nigripes*, $n = 5$), but only 2.25 μg g^{-1} dry wt in liver of the black-tailed gull (*Larus crassirostris*, $n = 5$) (see Fig. 7). The levels found in the black-tailed gull are comparable to those found in marine mammals (see Table 1). We also analyzed 24 additional liver samples of the black-footed albatross and found concentrations up to 42 μg g^{-1} dry wt (Fujihara et al. 2004), which is comparable to those in marine animals at lower trophic levels (Table 1). Arsenobetaine predominated in liver of both the black-footed albatross and black-tailed gull, with DMA(V), AC, and TETRA also being detected (Fig. 7; Kubota et al. 2003a). Interestingly, a low level of AB (0.19 μg g^{-1} dry wt) was observed in one liver sample from a jungle crow (*Corvus macrorhynchos*) specimen, although arsenic was not detected in four other specimens (Kubota et al. 2003a). Low levels of AB were also reported in some other terrestrial birds (Koch et al. 2005). The jungle crow might have obtained AB by eating leftover marine fish and shellfish from garbage. Alternatively, because some terrestrial organisms have recently been reported to contain AB at trace amounts (Kuehnelt and Goessler 2003), AB might have originated in terrestrial organisms consumed by the jungle crow.

Tissue distribution of arsenicals in avian species has, so far, been reported only for the black-tailed gull (Kubota et al. 2002b) and black-footed albatross (Fujihara et al. 2004). Among the 13 tissues analyzed from the black-tailed gull, the concentration of arsenic was highest in liver (mean, 1.59 μg g^{-1} dry wt), followed by kidney (mean, 1.17 μg g^{-1} dry wt), and was lowest in feathers (mean, 0.18 μg g^{-1} dry wt), with AB predominating in all the tissues (75%–97% of extractable arsenic; arsenic speciation was not conducted for feathers) (Kubota et al. 2002b). In contrast, the arsenic level was highest in lung (mean, 16 μg g^{-1} dry wt) and muscle (mean, 15 μg g^{-1} dry wt), and lowest in bone (mean, <0.1 μg g^{-1} dry wt) and feathers (mean, 0.70 μg g^{-1} dry wt), among the 17 tissues analyzed from the black-footed albatross (Fujihara et al. 2004). Similar to the black-footed albatross, marine animals at low trophic levels tend to accumulate arsenic in muscle. It is noteworthy that As(V) was detected in the muscle and testis of the black-footed albatross but that inorganic arsenic was not found in marine mammals.

The trophic transfer coefficient (TTC), defined as the ratio of the concentration in a consumer's body to the concentration in diet (stomach content) (Suedel et al. 1994), was found to be 1.0 for the black-footed albatross using arsenic levels of 17 tissues and diet (stomach content) (Fujihara et al. 2004). Because the TTC value is usually below unity for trace elements, other than those that are highly accumulative (e.g., mercury) (Anan et al. 2001), the black-footed albatross is believed to be very efficient in absorbing arsenic although arsenic does not biomagnify, thus leading to the levels reported (Fig. 8).

3.3 *Arsenic in Sea Turtles*

Few data are available on accumulation of arsenic or its species in sea turtles. In studies conducted in our laboratories (Saeki et al. 2000; Kubota et al. 2003a; Fujihara et al. 2003; Agusa et al. 2007), liver samples from the hawksbill turtle (*Eretmochelys imbricata*, $n = 19$) showed the highest arsenic levels (mean, 20.9 μg g^{-1} dry wt), followed by the loggerhead turtle (*Caretta caretta*, $n = 9$) (mean, 9.0 μg g^{-1} dry wt), and the lowest in the green turtle (*Chelonia mydas*, $n = 34$) (mean, 2.9 μg g^{-1} dry wt). Although the dugong, which feeds on seagrass, exhibited relatively high arsenic concentrations in the liver (see Fig. 6), the carnivorous species (i.e., hawksbill and loggerhead turtles) tended to show higher arsenic levels than did herbivorous sea turtles (i.e., green turtle). This pattern in sea turtles is similar to that observed in other low-trophic-level marine animals such as mollusks (Cullen and Reimer 1989).

Edmonds et al. (1994) described the first characterization of arsenic species in sea turtles. Arsenobetaine, As(III), and AC accounted for 50%, 35%, and 15% of water-extractable arsenic, respectively, in liver of the leatherback turtle (*Dermochelys coriacea*), whereas methanol extracts of AB and AC were 80% and 20%, respectively. The relatively high percentage of AC and As(III) is characteristic of the leatherback turtle, but this does not occur in marine mammals and seabirds. In studies conducted in our laboratories, AB predominated in liver of the green turtle, loggerhead turtle, and hawksbill turtle (see Fig. 7). Interestingly, the loggerhead turtle showed a relatively high percentage of AC (30% of extractable arsenic species) (Fig. 7), which was similar to that of the leatherback turtle

(Edmonds et al. 1994). It is known that most AC is converted to AB and also to a small amount of lipid-soluble arsenicals when administered to marine animals (Edmonds and Francesconi 2003). The high percentage of AC in these two turtle species may originate in their diets, because they feed primarily on jellyfish (Bjorndal 1997), some species of which are known to contain high AC levels (Hanaoka et al. 2001a). Alternatively, it is assumed that these sea turtles have a low capacity to convert AC to AB. It is noteworthy that green turtles feeding on marine algae and seagrass (Bjorndal 1997) have high proportions of AB in their livers (Fig. 7). Similar to the green turtle, the luderick (*Girella tricuspidata*), a herbivorous fish species, contained 67% AB and 15% DMA(V) of total arsenic extractable from the liver (Kirby and Maher 2002). Arsenosugars predominate in marine algae, which comprise the primary diet of the green turtle (Francesconi and Kuehnelt 2004). In our studies, although arsenosugars have not been measured, no significant HPLC peak other than AB, DMA(V), and AC has been detected in the HPLC-ICP-MS analysis for the green turtle (Kubota et al. 2002a, 2003a). It is known that arsenosugars absorbed through the diet are converted primarily to DMA(V) and are then excreted in the urine by humans (Le et al. 2004) and sheep (Martin et al. 2005). In the green turtle, however, the percentage of DMA(V) was low (Fig. 7), despite possible uptake by this species of large amounts of arsenosugars from marine algae. There are three possible explanations for these discrepancies: first, AB may be synthesized from arsenosugars by the green turtle or by intestinal bacteria they harbor; second, AB absorbed from diet animals (jellyfish and zooplankton) may be efficiently retained [adult green turtles feed on small amount of jellyfish, and zooplankton is the chief diet of juvenile turtles (Bjorndal 1997)], whereas DMA(V) converted from arsenosugars may be rapidly excreted; and, third, the green turtles might efficiently retain in the body any AB gleaned from marine algae, although only a small amount of AB may be present in marine algae (Nischwitz and Pergantis 2005a).

High concentrations of arsenic were observed in the liver (up to 32.8 μg/g dry wt) and muscle (205 μg/g) of hawksbill turtles (Saeki et al. 2000). These turtles may have a peculiar mechanism for arsenic accumulation, because their main food source, sponges, have rather low arsenic levels, when compared to other low-trophic-level marine organisms (Saeki et al. 2000). Fujihara et al. (2004) summarized the distribution of arsenic in tissues of various marine animals and concluded that species with high arsenic levels (e.g., hawksbill turtle and the black-footed albatross) tend to accumulate AB in the muscle. Such accumulation is also characteristic of some fish (Shiomi et al. 1996; Amlund et al. 2006a,b). Agusa et al. (2007), in reviewing the literature for arsenic levels in various marine animals, found the ratio of arsenic concentration in muscle versus liver to be high in sea turtles (5.87). Generally, inorganic arsenic is retained in mammalian tissues whereas organoarsenicals are rapidly excreted in urine (Shiomi 1994). In contrast, AB and AC tend to accumulate in fish tissues (especially muscle) while inorganic arsenic, DMA(V), and TMAO are readily excreted (Shiomi et al. 1996; Amlund et al. 2006b).

As(III) was detected at low levels in two of five green turtle liver samples and one of five loggerhead turtle liver samples (Kubota et al. 2003a). According to Agusa et al. (2007), As(III) was detected in all examined tissues of green and hawksbill turtles. Remarkably, high levels of As(III) were found in spleen of the hawksbill turtle (2.83 μg g^{-1} dry wt; Agusa et al. 2008). As(III) comprised 35% of water-extractable arsenic in the liver of the leatherback turtle (Edmonds et al. 1994). Storelli and Marcotrigiano (2000) analyzed organic and inorganic arsenic levels in the loggerhead turtle and found that inorganic arsenic comprised 3% and 11% of total arsenic in the muscle and liver, respectively. Although inorganic arsenic was not detected in liver of the hawksbill turtle by Fujihara et al. (2003), sea turtles generally have higher levels of inorganic arsenic than do marine mammals and seabirds.

A strong positive correlation was observed between AB and total arsenic concentrations in the liver of seabirds, sea turtles, and marine mammals (see Fig. 4). However, some differences exist in AB accumulation among species. Arsenobetaine was not detected in the dugong (not included in Fig. 4). Kubota et al. (2003a) reported that the proportion of AB increased with total arsenic concentration in marine mammals, seabirds, and sea turtles. Loggerhead turtles have a high arsenic level (mean, 11.2 μg g^{-1} dry wt; Fig. 7) and would, therefore, be expected to have high proportion of AB; however, the value was relatively low (mean, 54.0%). The low proportion of AB was attributed to high levels of AC in the loggerhead turtle (Fig. 7).

3.4 Toxicological Significance of Arsenic in Marine Mammals, Seabirds, and Sea Turtles

In general, inorganic arsenic is more toxic than organic arsenic (Shiomi 1994). Because AB is the dominant form in most marine mammals, seabirds, and sea turtles, risk to these marine animals may be rather low despite retention of high concentrations in their tissues. However, the more toxic inorganic form was detected in some specimens of sea turtles and seabirds. Inorganic arsenic acts as a carcinogen by forming certain reactive oxygen species (Kitchin 2001; Kitchin and Ahmad 2003; Hei and Filipic 2004). Oxidative damage to DNA is indeed reported in humans exposed to inorganic arsenic through contaminated groundwater (Feng et al. 2001; Basu et al. 2005; Kubota et al. 2006). However, oxidative stress induced by arsenic has not received much attention in marine organisms. Furthermore, arsenic has recently been accused of being a potent endocrine disruptor (Darbre 2006). Stoica et al. (2000) showed that As(III) activated the estrogen receptor-α (ER-α) through formation of a high-affinity complex with the hormone-binding domain of the receptor in human breast cancer cells. Bodwell et al. (2004, 2006) revealed that at very low levels As(III) stimulated transcription [mediated by glucocorticoid receptors (GR), progesterone receptors, and mineralocorticoid receptors of humans and rats), whereas at slightly higher but not cytotoxic concentrations, inhibition of transcription was observed. Waalkes et al. (2004) reported that exposure of inorganic arsenic can cause overexpression of ER-α through its promoter region hypomethylation in mice and humans. According to Stanton et al. (2006) and Shaw et al. (2007), inorganic arsenic may act as an endocrine disruptor in killifish; inorganic arsenic inhibits the ability of killifish to adapt to increased salinity by altering GR-mediated posttranscriptional steps that regulate cystic fibrosis transmembrane regulator (CFTR) protein abundance. Furthermore, some organoarsenicals such as DMA(V) and DMA(III), as well as inorganic arsenic, show carcinogenic action, probably by inducing oxidative stress (Kitchin 2001; Kitchin and Ahmad 2003; Hei and Filipic 2004). DMA(V) may be a carcinogen and tumor promoter in some experimental animals (Yamanaka et al. 2004). Presumably, dimethylarsenic peroxide may act as a tumor promoter and the dimethylarsenic radical and dimethylarsenic peroxy radical act as tumor-initiating factors, all of which seem to be metabolites of DMA(V) (Yamanaka et al. 2004). In HeLa S3 cells, As(III) induced oxidative DNA damage at $0.075\,\mu g\ ml^{-1}$, MA(III) and DMA(III) at $7.5\,\mu g\ ml^{-1}$, and MA(V) and DMA(V) at $750\,\mu g\ ml^{-1}$ (Schwerdtle et al. 2003). Because inorganic arsenic exerts adverse effects at low levels, its risk should be assessed in marine animals in which As(III) and As(V) are found. However, MA(V) and DMA(V) have not been found at levels that could adversely affect marine mammals, seabirds, and sea turtles (Fig. 7), but effects of their chronic exposure remain uncertain. Highly toxic MA(III) and DMA(III) have not been detected in marine organisms.

Arsenobetaine is known to be scarcely metabolized in animals, but small amounts of TMAO, TETRA, MA(V), DMA(V), As(V), and As(III) were detected in the urine of rats administered orally with AB (Yoshida et al. 2001). Excreted forms in the rat may result from degradation of AB by intestinal bacteria. Hence, effects of degradation products of AB, especially toxic inorganic arsenic, should be evaluated in marine animals known to absorb large amounts of AB from the organisms they consume.

Mammal species vary considerably in their capacity to methylate arsenic (Aposhian 1997; Vahter 1999); inorganic arsenic is methylated to MA(V) and DMA(V) in most mammalian species, but some species, such as the marmoset monkey and the chimpanzee, have low or no methylation capacity. In humans, genetic polymorphisms are known to affect arsenic biotransformation (Aposhian and Aposhian 2006), but no such information is available for marine organisms. For the future, information is needed on the metabolism of various forms of arsenic, not only in marine mammals, seabirds, and sea turtles but also in other marine animals and algae.

3.5 Newly Identified Arsenicals

Recently, various new arsenicals have been identified in tissues of marine mammals and other animals. Geiszinger et al. (2002) detected trimethylarsoniopropionate (TMAP) in muscle, liver, kidney, and lung of the sperm whale (*Physeter catodon*). The concentrations of TMAP were considerably lower than those of AB, but higher than those of DMA(V) and AC, and accounted for 3%–5% of total arsenicals found. Sloth et al. (2005a) detected DMAA, DMAP, and DMAE in the liver of hooded seal and DMAE in the kidney of harp seal. Mancini et al. (2006) identified a novel polyarsenic compound (arsenicin A; $C_3H_6As_4O_3$) in the marine sponge *Echinochalina bargibanti* from the coast of New Caledonia. Recently, various thio-arsenicals were also identified in the urine of sheep feeding on marine algae and humans exposed to arsenic from groundwater. In addition, 2-dimethylarsinothioyl acetic acid [$(CH_3)_2As(=S)CH_2COOH$] was detected in urine of wild sheep feeding on brown kelp (*Laminaria hyperborea*, *L. digitata*, etc.), which was the first identification of a thio-arsenical in mammals (Hansen et al. 2004a). Thio-dimethylarsinate [$(CH_3)_2As(=S)OH$; thio-DMA(V)] was also identified in urine of sheep (Hansen et al. 2003, 2004b). It is reported that humans convert arsenosugars to thio-arsenicals such as thio-DMAE and thio-DMAA (Raml et al. 2005). Furthermore, thio-DMA and thio-methylarsonate [$CH_3As(=S)(OH)_2$; thio-MA(V)] were identified in urine of humans exposed to inorganic arsenic in groundwater in Bangladesh (Raml et al. 2007). The sulfur of these thio-arsenicals is thought to be derived from H_2S, produced by sulfate-reducing bacteria in the gastrointestinal tract (Conklin et al. 2006) and released from cysteine degradation within cells (Hansen et al. 2004c). It is assumed that the oxo (As-O) and thio (As-S) forms have been readily interconverted (Raml et al. 2005). Surprisingly, Raab et al. (2007) identified a complex between thio-DMA(V) and glutathione in shoots of cabbage (*Brassica oleracea*) exposed to DMA(V), even though this is not a trivalent arsenic compound, suggesting that pentavalent arsinothioyl species may interact with proteins. We have not examined whether TMAP, DMAA, DMAP, and DMAE are present in other marine mammals, seabirds, and sea turtles because the corresponding standard compounds necessary for HPLC-ICP-MS analysis are unavailable. However, an unidentified arsenical, with behavior on HPLC similar to that of TMAP, was found in extracts from various marine mammals, seabirds, and sea turtles (Ebisuda et al. 2002; Kubota et al. 2003a, 2005). Also, thio-arsenicals, a new group of arsenic species, could be present in marine mammals, seabirds, and sea turtles, although they were not identified in the studies we conducted. We used ion-exchange columns for separation of arsenic species, but this method is not suitable for analysis of thio-arsenicals (Raml et al. 2006). Instead, a reverse-phase HPLC method would enable analysis of thio-arsenicals (Raml et al. 2006); thus, prospects for analyzing thio-arsenicals in marine mammals, seabirds, and sea turtles look promising.

4 Maternal Transfer of Arsenic Species

4.1 Maternal Transfer of Arsenic in Marine Mammals

Very few studies have been performed on the maternal transfer of arsenic species in marine mammals, seabirds, and sea turtles. Previously, it was reported that inorganic arsenic would pass through mammalian placentas but that organic arsenic would not (Morton and Dunnette 1994). However, women exposed to inorganic arsenic from drinking water contained mainly DMA(V) in cord blood, demonstrating placental transfer of some organoarsenicals in humans (Concha et al. 1998). Meador et al. (1993) detected arsenic in brain, liver, and kidney of fetal pilot whales, confirming placental transfer of arsenic in marine mammals. However, the arsenic species was not determined in either the mothers or fetuses. Kubota et al. (2005) studied maternal transfer of arsenic to the fetus of the Dall's porpoise. Analytes included liver, kidney, muscle, and blubber of both mother and fetus. Arsenobetaine, DMA(V), AC, and MA(V) were found in liver, kidney, and muscle of the fetus (arsenic speciation was not conducted in the blubber), and the arsenical composition was similar to that of the mother, suggesting that these arsenicals can transfer from mother to fetus. However, the arsenic level in the fetus was less than one-half of that in the mother. In the fetal Dall's porpoise, 25.2% and 59.0% of total arsenic burden was distributed in blubber and muscle, respectively, whereas 59.6% and 33.5% of the burden was distributed in blubber and muscle of the mother, respectively. This difference reflects the low placental transferability of lipid-soluble arsenicals from the mother's blubber (Ebisuda et al. 2002, 2003). It is reported that hydrophobic chemicals are much less transferable from mother to fetus in these marine mammals (Tanabe et al. 1982).

4.2 Maternal Transfer of Arsenic in Seabirds

Limited information is available on the maternal transfer of arsenicals to bird eggs. To our knowledge, only one study on arsenic species in bird eggs has been reported; DMA(V) and As(III) but not AB were detected in eggs of the spoonbill (*Platalea leucorodia*), although arsenic speciation was not conducted for the mother bird (Gómes-Ariza et al. 2000). Kubota et al. (2002b) studied the maternal transfer of arsenicals to eggs of the black-tailed gull. Arsenic composition in the eggs was similar to that in tissues of the mother bird, with AB being predominant, followed by DMA(V). However, the arsenic level in eggs was low compared to that in the mother bird. The eggs weighed 32% of the body weight of the mother black-tailed gull, but the percentage of arsenic in eggs was only 11% of that existing in the mother. For the black-tailed gull (Agusa et al. 2005), the transfer rate of arsenic from mother to eggs was comparable to that of vanadium, chromium, and antimony, which are generally less transferable in birds. Arsenic was detected in eggs of sea turtles (Lam et al. 2006), and, thus, is confirmed to transfer to eggs. However, as far as we know, no studies exist on the nature of arsenic species in sea turtle eggs.

5 Arsenobetaine: Accumulation Mechanism and Origin

5.1 Origin and Synthetic Pathway of Arsenobetaine

The origin and synthetic pathways followed by AB are controversial. Principally, four pathways for synthesis of AB have been proposed (Fig. 9). In the first two, AB is transformed from dimethylated arsenosugars through DMAA or AC (Fig. 9a,b).

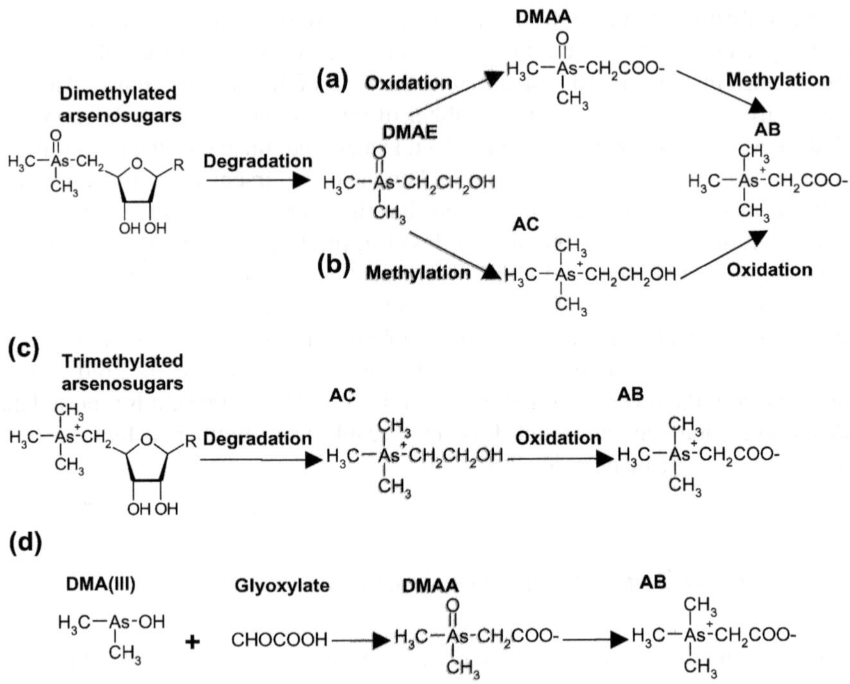

Fig. 9 Hypothesized synthetic pathways of arsenobetaine

In the third, AB is converted from trimethylated arsenosugars (Fig. 9c). Finally, it is postulated that AB is synthesized from DMA(III) and 2-oxo acids, glyoxylate (Fig. 9d) and pyruvate, a similar pathway as exists for amino acid biosynthesis. DMAE and DMAA, at low levels, were observed in several marine animals and marine algae (Sloth et al. 2005a). The existence of these two forms supports the concept that a pathway exists from dimethylated arsenosugars (Fig. 9a,b) to AB. Despite the natural occurrence of trimethylated arsenosugars in marine organisms, their very low concentrations do not account for the presence of AB at a high concentration in marine animals (Edmonds and Francesconi 2003). However, relatively high concentrations of a trimethylated arsenosugar (2'3'-dihydroxypropyl 5-deoxy-5-trimethyl arsonioriboside) were detected in abalone (*Haliotis rubra*) from New South Wales, Australia. This arsenosugar accounted for 28% (5 $\mu g\ g^{-1}$ dry wt) of all arsenicals in intestinal tissue and 0.9% (0.4 $\mu g\ g^{-1}$ dry wt) of the total in muscle (Kirby et al. 2005). Hence, the trimethylated arsenosugar might contribute to synthesis of AB in this marine animal. The pathway that produces AB in deep-sea organisms and some terrestrial ones that are not dependent on marine algae is unclear (Edmonds and Francesconi 2003). A pathway (Fig. 9d) starting from DMA(III), recently proposed by Edmonds (2000), could explain the presence of AB in these animals and also some other arsenicals found in marine organisms (Edmonds and Francesconi 2003). For example, DMAA could be synthesized in this pathway, although DMAE, an arsenical occasionally found in marine organisms, is not part of this pathway. Furthermore, DMAP and TMAP, which are found in various marine animals and algae, could be synthesized from oxaloacetate instead of glyoxylate (Fig. 9d). The major pathway leading to production of AB may vary with location, organism, and ecosystem. To be sure, a comprehensive understanding of this topic requires further research studies.

Arsenobetaine has been detected in marine animals but not in marine algae (Francesconi and Edmonds 1993), so it is presumed that synthesis of AB is ascribed to metabolic capacities of the animals and their intestinal bacteria. Involvement of bacteria in AB synthesis has been reported by several researchers. Ritchie et al. (2004) observed that AB was synthesized from DMAA and the methyl donor *S*-adenosylmethionine by lysed-cell extracts of *Pseudomonas fluorescens* A (NCIMB 13944) isolated from the blue mussel. Interestingly, this bacterium was known to degrade AB to DMAA (Jenkins et al. 2003), so the reaction in both directions might be catalyzed by a methyltransferase (Ritchie et al. 2004). The results of Ritchie et al. (2004) also suggest a direct involvement by bacteria in synthesis of AB within marine animals. In 2005, Nischwitz and Pergantis (2005a) identified and quantified AB in marine algae, by HPLC-ES-MS/MS, for the first time. In this study, commercially available brown algal powder, and fresh green, brown, and red algae, which were carefully washed to remove epifauna and contaminants from the surface, were employed; visible epifauna with body size >0.1 mm were fully removed, especially from transparent green algae. Furthermore, extraction was performed by a mild procedure using only deionized water, and methanol and sonication were avoided to prevent arsenic transformation. Analysis under these conditions revealed that AB accounted for 7.5% of extracted arsenic in green algae and 0.25%–1.3% in other algal samples. These authors also indicate that the chromatographic peak of AB present in trace amounts cannot be separated chromatographically from the larger peaks of the major arsenicals (i.e., arsenosugars) in marine algae, and therefore the presence of AB could not be confirmed in marine algae by HPLC-ICP-MS analysis (Nischwitz and Pergantis 2005a). It should be noted that the HPLC-ES-MS/MS enables analysis of arsenicals co-eluting from the HPLC column, in contrast to HPLC-ICP-MS. These results cast doubt on the general assumption that AB is not present in marine algae and the belief that this arsenical is either synthesized or accumulated only in marine animals. Therefore, further studies are needed to confirm these findings.

5.2 *Accumulation Mechanism of Arsenobetaine in Marine Animals*

It has been pointed out that high levels of AB in marine animals may be related to the salinity of seawater. Organisms are known to utilize various osmolytes, low molecular weight osmotically active solutes, to adapt to osmotic stress. Glycine betaine [GB, $(CH_3)_3N^+CH_2COO^-$] is the nitrogen analogue of AB (Shibata et al. 1992) and behaves as an osmolyte in marine animals at lower trophic levels (Yancey et al. 1982), in mammals (Burg et al. 1997), and in birds (Lien et al. 1993). It is suggested that once AB is synthesized in the marine food chain, it may be taken up into cells through the same route that absorbs GB and thereby behave as GB does in the cells (Shibata et al. 1992; Shibata and Morita 2000). Indeed, it has been reported that neither GB nor AB is bound to macromolecules (e.g., proteins) (Vahter et al. 1983).

Arsenobetaine was not detected in the bivalve, *Corbicula japonica,* which lives in a low-salinity estuary (Shibata and Morita 1992). This result suggests the possibility that AB was not accumulated in the bivalve because osmolytes are not necessary in a low-salinity environment. It was also shown that the blue mussel accumulated AB efficiently from seawater, but the accumulation decreased in the presence of GB in seawater (Gailer et al. 1995). Clowes and Francesconi (2004) reported that AB levels increased in the blue mussel when the animals were maintained at high salinity. When the blue mussel that had been maintained at high salinity was transferred to low-salinity seawater, AB levels decreased in the gill but not in other tissues (Clowes and Francesconi 2004). These results suggest that AB behaves as an osmolyte in the blue mussel and that the gill responds sooner to osmotic changes than did other tissues. Also, for herring (*Clupea harengus*), cod (*Gadus morhua*), and flounder (*Platichthys flesus*), total arsenic concentrations in muscle were correlated with salinity at locations where the fish were collected, which may be because arsenic levels (probably AB) in fish or their diet animals reflected the ambient salinity (Larsen and Francesconi 2003). Although controversial, Amlund and Berntssen (2004), after studying the retention capacity of AB in seawater- and freshwater-adapted Atlantic salmon (*Salmo salar*), found no significant difference between such groups, despite the AB level in muscle of seawater-adapted wild salmon being more than 10 fold that of freshwater-adapted wild salmon. Thus, the high AB level of seawater-adapted wild salmon might be caused by the AB level in diet rather than an adaptation to salinity. On the other hand, the bacterium *Escherichia coli* (Pichereau et al. 1997) and Madin Darby canine kidney (MDCK) cells (Randall et al. 1996) absorbed AB and GB in response to osmotic stress, although the uptake rate of AB was lower than that of GB (Randall et al. 1996). Furthermore, it was shown that AB was efficiently absorbed through two GB transporters, ProP and ProU, in *Escherichia coli* (Randall et al. 1995). An alternative explanation for high concentrations of AB in marine animals is that AB might be largely distributed in cellular organelles. Vahter et al. (1983) reported that urinary excretion of AB was slower in rabbits, which have more AB in cellular organelles, than in mice or rats. The relationship between AB accumulation and its subcellular distribution has not yet been examined in wildlife.

To gain insight into the mechanisms of the high AB accumulation, we determined subcellular distribution of arsenic and the relationship between AB and GB concentrations in livers of the northern fur seal (*Callorhinus ursinus*), ringed seal, black-footed albatross, black-tailed gull, hawksbill turtle, and green turtle (Fujihara et al. 2003). Results indicated that arsenic levels were not related to the subcellular distribution in these marine animals. However, a significant negative correlation was observed between AB and GB concentrations for all animals examined (Fig. 10a); a strong negative correlation was observed, especially for the black-footed albatross

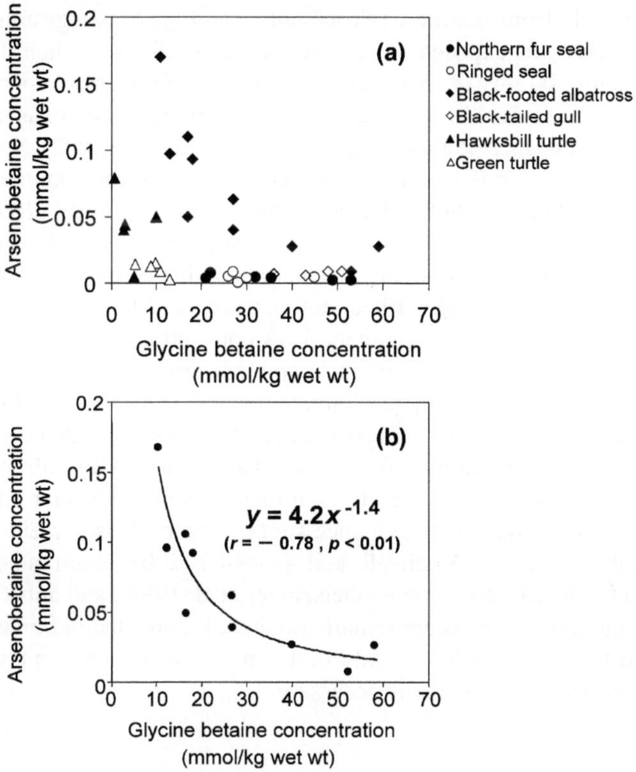

Fig. 10 Relationship between arsenobetaine and glycine betaine concentrations in liver of marine mammals, seabirds, and sea turtles (Fujihara et al. 2003)

(Fig. 10b). These results were contrary to our expectation; we had assumed that GB might increase with increasing AB levels in the animals, but, instead, a negative correlation was observed (Fig. 10). It is assumed that AB and GB are both taken up and efficiently retained when osmolyte content (e.g., GB) is insufficient in these animals or in their food supply. If true, this condition would lead to a negative correlation between AB and GB levels in the marine animals. It is likely that the contribution of AB to osmoregulation was low because the AB level was remarkably lower than that of GB (Fig. 10).

5.3 *Arsenobetaine in Freshwater and Terrestrial Environments*

Arsenic concentrations in seals from freshwater environments, the Baikal seal (*Pusa sibirica*) (Kubota et al. 2001) from Russia, and the harbor seal (*Phoca vitulina*) from northern Quebec in Canada (Langlois and Langis 1995), were lower than those in seals from marine environments (see Fig. 6), suggesting that these freshwater species are exposed to low AB concentrations in their food supply because of the low salinity of freshwater habitats. It is frequently reported that AB is not a dominant arsenical in freshwater animals, although arsenosugars dominate in both freshwater and marine algae (Francesconi and Kuehnelt 2002). Nondetectable or very low concentrations of AB was observed in freshwater animals from the River Danube in Hungary (Schaeffer et al. 2006). Jankong et al. (2007) reported relatively low accumulation of AB in tissues, especially liver, of four species of freshwater fish collected from highly arsenic-contaminated ponds (550 and 990 μg L^{-1}). Soeroes et al. (2005) described the absence of AB in the common carp (*Cyprinus carpio*) from lakes in Hungary. Low concentrations of AB in freshwater fish may reflect the low salinity of their ambient environment. However, some freshwater fish species contain predominantly AB (Shibata and Morita 2000; Francesconi and Kuehnelt 2002). Šlejkovec et al. (2004) suggest that composition of arsenic species is different among freshwater fish species; AB predominates especially in species of salmonids even though they inhabit freshwater environments all through their life. Increasing evidence suggests that AB is present also in various terrestrial organisms (Kuehnelt and Goessler 2003) such as mushrooms (Kuehnelt et al. 1997a), earthworms (Geiszinger et al. 1998), and ants (Kuehnelt et al. 1997b), although the levels are remarkably low. Future elucidation of the origin, behavior, and function of AB, not only in the marine ecosystem, but also in freshwater and terrestrial ecosystems, is necessary.

6 Lipid-Soluble Arsenic

6.1 *Lipid-Soluble Arsenicals in Marine Organisms at Low Trophic Levels*

It is well known that various marine organisms contain lipid-soluble arsenic. The ascidian (*Halocynthia roretzi*), the turban shell (*Turbo cornutus*), the short-necked clam (*Tapes japonica*) (Shinagawa et al. 1983), the red sea urchin (*Pseudocentrotus depressus*), the abalone (*Haliotis diversicolor supertexta*), the three-line grunt (*Parapristipoma trilineatum*), and the Japanese surfperch (*Neoditrema ransonneti*) (Kaise et al. 1988) all showed relatively high concentrations of lipid-soluble arsenic, although the levels were lower than those of water-soluble arsenic. A phosphatidylarsenosugar was identified for the first time in a brown alga (*Undaria pinnatifida*) as a lipid-soluble arsenical in 1988 (Fig. 11; Morita and Shibata 1988). Phosphatidylarsenocholine (Fig. 11), a phosphatidylcholine analogue, was also identified in the digestive gland of the Western rock lobster (Edmonds et al. 1992). Because direct introduction of organic solvents, used for extraction of lipid-soluble arsenicals, into ICP-MS is generally problematic, lipid-soluble arsenicals in marine organisms are poorly studied. Until now, only water-soluble arsenicals

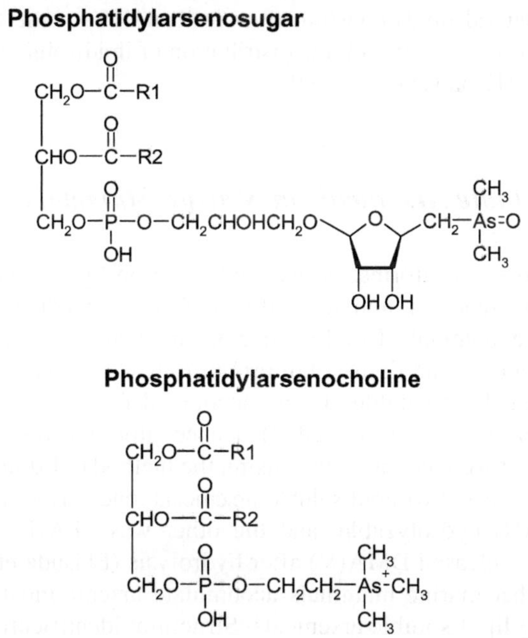

Fig. 11 Lipid-soluble arsenicals identified in marine organisms

released from the lipid-soluble arsenicals have been studied after alkaline or acid digestion. Unfortunately, these analyses cannot provide direct information on the structure of lipid-soluble arsenicals.

In the star spotted shark (*Mustelus manazo*), almost all lipid-soluble arsenic was detected in the polar lipid fraction. Lipid-soluble arsenic levels in this fraction were particularly high in liver, gallbladder, and kidney (Hanaoka et al. 1999). Lipid-soluble arsenic in the shark was fractionated into alkali (0.027 M NaOH)-stable and alkali-labile portions, with the former predominating in liver and gallbladder, and the latter in kidney (Hanaoka et al. 1999). Thus, the results suggest that the nature of lipid-soluble arsenicals varies among tissues of the shark. Altogether, four arsenicals were detected in the alkali-labile fraction of 12 shark tissues, with one of these arsenicals predominating in kidney, spleen, and brain (Hanaoka et al. 2001b). After digestion with 6 M HCl of three arsenicals detected in the alkali-labile fraction, AC or DMA(V) was released from two of three of these arsenicals; HCl failed to digest the third arsenical. Hydrolysis of the alkali-stable fraction with saturated $Ba(OH)_2$ gave primarily DMA(V) for liver, and DMA(V) and AC for muscle, skin, stomach, and intestine (Hanaoka et al. 2001b). Therefore, at least six lipid-soluble arsenicals exist in and contain precursors of DMA and AC in the star spotted shark. In contrast, water-soluble arsenicals were not released by alkaline hydrolysis (1 M NaOH), but DMA(V) was detected after acid digestion (conc. HNO_3) in fish oil (Kohlmeyer et al. 2005). Lipid-soluble arsenicals were only detected in the polar lipid fraction of fish oil, where neutral lipids constituted more than 90% of the total (Kohlmeyer et al. 2005); this is consistent with the distribution of lipid-soluble arsenicals in the star spotted shark (Hanaoka et al. 1999).

6.2 Lipid-Soluble Arsenicals in Marine Mammals

As far as we know, no information was available on lipid-soluble arsenicals in marine mammals, seabirds, and sea turtles before we conducted studies on the blubber of marine mammals. Lipid-soluble arsenic was found in ringed seal liver, kidney, muscle, and gonad tissues, but proportions were very low (Ebisuda et al. 2002). In contrast, lipid-soluble arsenic accounted for about 90% of arsenic in ring seal blubber (Ebisuda et al. 2002). Hence, lipid-soluble arsenicals in the ringed seal blubber were characterized using the method of Edmonds et al. (1992). Results showed at least two lipid-soluble arsenicals: one was tetraethylammonium hydroxide (TEAH) hydrolyzable, and the other was TEAH stable but NaOH labile. Both these released DMA(V) after hydrolysis (Ebisuda et al. 2003). Thus, it is suggested that marine mammals accumulate arsenic mostly in blubber as DMA-containing lipid-soluble arsenicals. Structural identification of these lipid-soluble arsenicals will be critical to understanding how arsenic is metabolized in marine mammals. Insights into how DMA(V) is incorporated into lipid-soluble arsenicals in marine animals would also be useful, considering the genotoxic potential of DMA(V).

6.3 Promising Analytical Methods for Lipid-Soluble Arsenicals

Miyajima et al. (1988) purified and analyzed, with nuclear magnetic resonance (NMR), a lipid-soluble arsenical from the tiger shark (*Galeocerdo cuvier*) without digestion that revealed the presence of $(CH_3)_2As(O)CH_2^-$ in this arsenical. Devalla and Feldmann (2003) characterized lipid-soluble arsenicals by an enzymatic hydrolytic method. Water-soluble arsenicals, released by phospholipase D treatment of lipid-soluble ones from a marine alga, was mainly an arsenosugar. In contrast, DMA(V) and MA(V) were the major arsenicals released from lipid-soluble arsenicals in kidney, and muscle, and DMA(V) from lipid-soluble ones in feces of marine algae-eating sheep. Also, the digestion of the lipid-soluble arsenicals by phospholipase D indicated that the released arsenicals were bound not to the simple lipid (comprising 91% of total lipid) but to a complex lipid (e.g., phospholipid and sphingolipid) present in sheep tissues. Importantly, limited numbers of water-soluble arsenicals released from lipid-soluble arsenicals may result from binding of an arsenical moiety to various lipid-soluble species (Schmeisser et al. 2006a). Ninh et al. (2007) suggested the possible presence of two DMA(V)-containing lipid-soluble arsenicals, phosphatidyldimethylarsinic acid and DMA(V)-containing sphingomyelin, in the

Japanese flying squid (*Todarodes pacificus*) using a combination of chemical and enzymatic hydrolysis techniques.

Little is known about the toxicity of lipid-soluble arsenicals in organisms (Francesconi 2005). In humans, after metabolism, DMA(V) and trace amounts of oxo-DMAP, thio-DMAP, oxo-DMAB, and thio-DMAB were excreted in the urine (Schmeisser et al. 2006a,b). The toxicological evaluation of lipid-soluble arsenicals and their metabolites will be the subject of future research.

Because it has been difficult to directly analyze for lipid-soluble arsenicals by HPLC-ICP-MS, very little progress has been made with lipid-soluble arsenicals in contrast to water-soluble arsenicals. Very recently, it became feasible to analyze for lipid-soluble arsenicals directly by HPLC-ICP-MS (Schmeisser et al. 2005). These authors found that problems associated with the introduction of organic solvent to the plasma were considerably reduced by using a low column flow rate, a cooled spray chamber (−5°C), and addition of oxygen directly to the plasma (20% in argon). The authors applied this method to successfully observe the presence of at least 10 lipid-soluble arsenicals in fish oil. This method would be useful to characterize lipid-soluble arsenicals in marine mammals, seabirds, and sea turtles.

7 Future Areas of Study

More than 30 arsenicals have been identified in marine environments (Francesconi and Kuehnelt 2004), and it is expected that more will be identified as analytical techniques advance. For example, 15 unknown trace arsenicals, which were not detected by conventional methods such as HPLC-ICP-MS and HPLC-hydride generation (HG)-ICP-MS, were found in human urine using LC-high-efficiency photooxidation (HEPO)-HG-ICP-MS (Nakazato and Tao 2006). The development of such new analytical techniques is important because such tools improve our understanding of arsenics behavior in geochemical cycles, ecosystems, organisms, and transformation pathways, as well as their toxicity. If progress is to be made, convenient methods of synthesis for standard compounds of recently identified arsenicals and commercial production thereof are badly needed. As recently identified arsenicals (e.g., DMAA and TMAP) have been quantified in some CRMs such as DORM2 (dogfish muscle) and BCR CRM627 (tuna fish tissue) (Sloth et al. 2003), these CRMs can be utilized to evaluate the accuracy of methods for these arsenicals.

Studies on the interaction between arsenic and other elements in marine organisms are also necessary. It is well known that mercury is detoxified through binding to selenium and/or sulfur in marine mammals and seabirds (Shibata et al. 1992; Ng et al. 2001; Arai et al. 2004; Ikemoto et al. 2004). Similarly, interaction between arsenic and selenium and/or sulfur may also occur (Gailer 2007). A metabolite containing glutathione (GSH) and equimolar amounts of As and Se, $[(GS)_2AsSe]^-$, was identified in bile from rabbits injected with selenite [Se(IV)] and As(III) (Gailer et al. 2000). This metabolite is thought to be synthesized in hepatocytes (Gailer et al. 2002a) and erythrocytes (Manley et al. 2006) with high endogenous concentrations of GSH. Furthermore, $[(CH_3)_2AsSe_2]^-$ can be synthesized chemically by reacting DMA(V) with GSH and Se(IV) (Gailer et al. 2002b). Even though these metabolites are present in marine mammals, seabirds, and sea turtles, the levels may be low. Small granules (diameter about 3 nm) of As_2Se were observed in the kidney of rats injected with inorganic arsenic and selenium (Berry and Galle 1994), and thus it will be interesting to examine whether such granules are present in marine animals. Also, Kanaki and Pergantis (2007) recently synthesized seleno-arsenosugars and seleno-DMA(V) by reacting arsenosugars and DMA(V) with H_2Se, respectively; these reaction products are unstable and, if present, are not expected to occur in organisms at significant levels.

DMA(V), which is widely distributed in marine animals, was thought to be a metabolite formed during the detoxification of inorganic arsenic; however, the metabolites of the methylation process for inorganic arsenic (see Fig. 3) include MA(III) and DMA(III), which have recently been reported to be highly toxic; hence, this process is no longer regarded as one of detoxification for inorganic arsenic. The possible carcinogenicity of DMA(V) should induce additional interest in conducting safety evaluations on marine animals. Furthermore, arsenosugars, which are generally considered to be practically nontoxic, exhibit genotoxicity by forming reactive oxygen species when present as trivalent arsenicals (Andrewes et al. 2004). Such trivalent arsenosugars are readily formed by the reaction of pentavalent arsenosugars with thiols (Andrewes et al. 2004). Additional evaluations of the *in vivo* toxicity of arsenosugars are therefore needed.

Nonextractable arsenic in the tissues of marine organisms should be studied. For example, an average of 0.73 μg g^{-1} dry wt of arsenic was present as nonextractable residue after extraction with methanol:water (9:1 v/v) in the liver of green turtles (Kubota et al. 2003a). This residue constituted about 25% of total arsenic. Thus, nonextractable arsenic cannot be ignored if the complete pattern of arsenic metabolism in marine organisms is to be understood. Arsenic-binding proteins are known to exist in experimental animals, and such proteins may be involved in the transformation and detoxification of arsenic (Aposhian and Aposhian 2006). For example, it was reported that most of the As(III) added to rat liver cytosol became protein bound, and this binding affected the extent of its subsequent methylation (Styblo et al. 1996; Styblo and Thomas 1997). Trivalent methylated arsenicals, MA(III) and DMA(III), also bind to proteins (Naranmandura et al. 2006). Further work is needed to clarify the role arsenic-bound proteins play in the metabolism of arsenic by marine mammals, seabirds, and sea turtles.

8 Summary

Although there have been numerous studies on arsenic in low-trophic-level marine organisms, few studies exist on arsenic in marine mammals, seabirds, and sea turtles. Studies on arsenic species and their concentrations in these animals are needed to evaluate their possible health effects and to deepen our understanding of how arsenic behaves and cycles in marine ecosystems. Most arsenic in the livers of marine mammals, seabirds, and sea turtles is AB, but this form is absent or occurs at surprisingly low levels in the dugong. Although arsenic levels were low in marine mammals, some seabirds, and some sea turtles, the black-footed albatross and hawksbill and loggerhead turtles showed high concentrations, comparable to those in marine organisms at low trophic levels. Hence, these animals may have a specific mechanism for accumulating arsenic. Osmoregulation in these animals may play a role in the high accumulation of AB. Highly toxic inorganic arsenic is found in some seabirds and sea turtles, and some evidence suggests it may act as an endocrine disruptor, requiring new and more detailed studies for confirmation. Furthermore, DMA(V) and arsenosugars, which are commonly found in marine animals and marine algae, respectively, might pose risks to highly exposed animals because of their tendency to form reactive oxygen species. In marine mammals, arsenic is thought to be mainly stored in blubber as lipid-soluble arsenicals. Because marine mammals occupy the top levels of their food chain, work to characterize the lipid-soluble arsenicals and how they cycle in marine ecosystems is needed. These lipid-soluble arsenicals have DMA precursors, the exact structures of which remain to be determined. Because many more arsenicals are assumed to be present in the marine environment, further advances in analytical capabilities can and will provide useful future information on the transformation and cycling of arsenic in the marine environment.

Acknowledgments We are grateful to Dr. Y. Shibata (National Institute for Environmental Studies, Japan), Prof. N. Miyazaki (the University of Tokyo, Japan), Dr. J. Yang (Freshwater Fisheries Research Center, China), Prof. H. Ogi (Hokkaido University, Japan), Dr. H. Tanaka (National Research Institute of Fisheries and Environment of Inland Sea, Japan), and Mr. K. Ebisuda (Ehime University, Japan) for their help and discussion. We also thank Prof. A. Subramanian (Ehime University, Japan) for critical reading of the manuscript. Work on these topics in our laboratory was supported by the "21st Century COE Program" and "Global COE Program" from the Ministry of Education, Culture, Sports, Science and Technology (MEXT), Japan and Japan Society for the Promotion of Science (JSPS), respectively.

References

Agusa T, Matsumoto T, Ikemoto T, Anan Y, Kubota R, Yasunaga G, Kunito T, Tanabe S, Ogi H, Shibata Y (2005) Body distribution of trace elements in black-tailed gulls from Rishiri Island, Japan: age-dependent accumulation and transfer to feathers and eggs. Environ Toxicol Chem 24:2107–2120.

Agusa T, Takagi K, Kubota R, Anan Y, Iwata H, Tanabe S (2008) Specific accumulation of arsenic compounds in green turtles (*Chelonia mydas*) and hawksbill turtles (*Eretmochelys imbricata*) from Ishigaki Island, Japan. Environ Pollut (in press).

Amlund H, Berntssen MHG (2004) Arsenobetaine in Atlantic salmon (*Salmo salar* L.): influence of seawater adaptation. Comp Biochem Physiol Part C 138:507–514.

Amlund H, Ingebrigtsen K, Hylland K, Ruus A, Eriksen DØ, Berntssen MHG (2006a) Disposition of arsenobetaine in two marine fish species following administration of a single oral dose of [^{14}C]arsenobetaine. Comp Biochem Physiol Part C 143:171–178.

Amlund H, Francesconi KA, Bethune C, Lundebye A-K, Berntssen MHG (2006b) Accumulation and elimination of dietary arsenobetaine in two species of fish, Atlantic salmon (*Salmo salar* L.) and Atlantic cod (*Gadus morhua* L.). Environ Toxicol Chem 25:1787–1794.

Anan Y, Kunito T, Watanabe I, Sakai H, Tanabe S (2001) Trace element accumulation in hawksbill turtle (*Eretmochelys imbricata*) and green turtle (*Chelonia mydas*) from Yaeyama Islands, Japan. Environ Toxicol Chem 20:2802–2814.

Andrewes P, Demarini DM, Funasaka K, Wallace K, Lai VWM, Sun H, Cullen WR, Kitchin KT (2004) Do arsenosugars pose a risk to human health? The comparative toxicities of a trivalent and pentavalent arsenosugar. Environ Sci Technol 38:4140–4148.

Aposhian HV (1997) Enzymatic methylation of arsenic species and other new approaches to arsenic toxicity. Annu Rev Pharmacol Toxicol 37:397–419.

Aposhian HV, Aposhian MM (2006) Arsenic toxicology: five questions. Chem Res Toxicol 19:1–15.

Arai T, Ikemoto T, Hokura A, Terada Y, Kunito T, Tanabe S, Nakai I (2004) Chemical forms of mercury and cadmium accumulated in marine mammals and seabirds as determined by XAFS analysis. Environ Sci Technol 38:6468–6474.

Azcue JM, Nriagu JO (1994) Arsenic: historical perspectives. In: Nriagu JO (ed) Arsenic in the Environment, Part I: Cycling and Characterization. Wiley, New York, pp 1–15.

Basu A, Som A, Ghoshal S, Mondal L, Chaubey RC, Bhilwade HN, Rahman MM, Giri AK (2005) Assessment of DNA damage in peripheral blood lymphocytes of individuals susceptible to arsenic induced toxicity in West Bengal, India. Toxicol Lett 159:100–112.

Berry JP, Galle P (1994) Selenium–arsenic interaction in renal cells: role of lysosomes. Electron microprobe study. J Submicrosc Cytol Pathol 26:203–210.

Bjorndal KA (1997) Foraging ecology and nutrition of sea turtles. In: Lutz PL, Musick JA (eds) The Biology of Sea Turtles. CRC Press, Boca Raton, FL, pp 199–231.

Bodwell JE, Kingsley LA, Hamilton JW (2004) Arsenic at very low concentrations alters glucocorticoid receptor (GR)-mediated gene activation but not GR-mediated gene repression: complex dose–response effects are closely correlated with levels of activated GR and require a functional GR DNA binding domain. Chem Res Toxicol 17:1064–1076.

Bodwell JE, Gosse JA, Nomikos AP, Hamilton JW (2006) Arsenic disruption of steroid receptor gene activation: complex dose–response effects are shared by several steroid receptors. Chem Res Toxicol 19:1619–1629.

Braune BM, Outridge PM, Fisk AT, Muir DCG, Helm PA, Hobbs K, Hoekstra PF, Kuzyk ZA, Kwan M, Letcher RJ, Lockhart WL, Norstrom RJ, Stern GA, Stirling I (2005) Persistent organic pollutants and mercury in marine biota of the Canadian Arctic: an overview of spatial and temporal trends. Sci Total Environ 351–352:4–56.

Burg MB, Kwon ED, Kültz D (1997) Regulation of gene expression by hypertonicity. Annu Rev Physiol 59:437–455.

Chowdhury AMR (2004) Arsenic crisis in Bangladesh. Sci Am 291:86–91.

Clowes LA, Francesconi KA (2004) Uptake and elimination of arsenobetaine by the mussel *Mytilus edulis* is related to salinity. Comp Biochem Physiol Part C 137:35–42.

Concha G, Vogler G, Lezcano D, Nermell B, Vahter M (1998) Exposure to inorganic arsenic metabolites during early human development. Toxicol Sci 44:185–190.

Conklin SD, Ackerman AH, Fricke MW, Creed PA, Creed JT, Kohan MC, Herbin-Davis K, Thomas DJ (2006) *In vitro* biotransformation of an arsenosugar by mouse anaerobic cecal microflora and cecal tissue as examined using IC-ICP-MS and LC-ESI-MS/MS. Analyst 131:648–655.

Cox PA (1995) The Elements on Earth: Inorganic Chemistry in the Environment. Oxford University Press, Oxford, pp 287.

Cullen WR, Reimer KJ (1989) Arsenic speciation in the environment. Chem Rev 89:713–764.

Darbre PD (2006) Metalloestrogens: an emerging class of inorganic xenoestrogens with potential to add to the oestrogenic burden of the human breast. J Appl Toxicol 26:191–197.

Devalla S, Feldmann J (2003) Determination of lipid-soluble arsenic species in seaweed-eating sheep from Orkney. Appl Organomet Chem 17:906–912.

Devesa V, Loos A, Súñer MA, Vélez D, Feria A, Martínez A, Montoro R, Sanz Y (2005) Transformation of organoarsenical species by the microflora of freshwater crayfish. J Agric Food Chem 53:10297–10305.

Ebisuda K, Kunito T, Kubota R, Tanabe S (2002) Arsenic concentrations and speciation in the tissues of the ringed seals (*Phoca hispida*) from Pangnirtung, Canada. Appl Organomet Chem 16:451–457.

Ebisuda K, Kunito T, Fujihara J, Kubota R, Shibata Y, Tanabe S (2003) Lipid-soluble and water-soluble arsenic compounds in blubber of ringed seal (*Pusa hispida*). Talanta 61:779–787.

Edmonds JS (2000) Diastereoisomers of an 'arsenomethionine'-based structure from *Sargassum lacerifolium*: the formation of the arsenic-carbon bond in arsenic-containing natural products. Bioorg Med Chem Lett 10:1105–1108.

Edmonds JS, Francesconi KA (1981) Arseno-sugars from brown kelp (*Ecklonia radiata*) as intermediates in cycling of arsenic in a marine ecosystem. Nature (Lond) 289:602–604.

Edmonds JS, Francesconi KA (2003) Organoarsenic compounds in the marine environment. In: Craig PJ (ed) Organometallic Compounds in the Environment. Wiley, New York, pp 195–222.

Edmonds JS, Francesconi KA, Cannon JR, Raston CL, Skelton BW, White AH (1977) Isolation, crystal structure and synthesis of arsenobetaine, the arsenical constituent of the western lock lobster *Panulirus longipes cygnus* George. Tetrahedron Lett 18:1543–1546.

Edmonds JS, Shibata Y, Francesconi KA, Yoshinaga J, Morita M (1992) Arsenic lipids in the digestive gland of the western rock lobster *Panulirus cygnus*: an investigation by HPLC ICP-MS. Sci Total Environ 122:321–335.

Edmonds JS, Francesconi KA, Stick RV (1993) Arsenic compounds from marine organisms. Nat Prod Rep 10:421–428.

Edmonds JS, Shibata Y, Prince RIT, Francesconi KA, Morita M (1994) Arsenic compounds in tissues of the leatherback turtle, *Dermochelys coriacea*. J Mar Biol Assoc U K 74:463–466.

Edmonds JS, Shibata Y, Francesconi KA, Rippingale RJ, Morita M (1997) Arsenic transformations in short marine food chains studied by HPLC-ICP MS. Appl Organomet Chem 11:281–287.

Eisler R (1994) A review of arsenic hazards to plants and animals with emphasis on fishery and wildlife resources. In: Nriagu JO (ed) Arsenic in the Environment, Part II: Human Health and Ecosystem Effects. Wiley, New York, pp 185–259.

Feng Z, Xia Y, Tian D, Wu K, Schmitt M, Kwok RK, Mumford JL (2001) DNA damage in buccal epithelial cells from individuals chronically exposed to arsenic via drinking water in Inner Mongolia, China. Anticancer Res 21:51–58.

Francesconi KA (2005) Current perspectives in arsenic environmental and biological research. Environ Chem 2:141–145.

Francesconi KA, Edmonds JS (1993) Arsenic in the sea. Oceanogr Mar Biol Annu Rev 31:111–151.

Francesconi KA, Kuehnelt D (2002) Arsenic compounds in the environment. In: Frankenberger WT Jr (ed) Environmental Chemistry of Arsenic. Dekker, New York, pp 51–94.

Francesconi KA, Kuehnelt D (2004) Determination of arsenic species: a critical review of methods and applications, 2000–2003. Analyst 129:373–395.

Fricke MW, Creed PA, Parks AN, Shoemaker JA, Schwegel CA, Creed JT (2004) Extraction and detection of a new arsine sulfide containing arsenosugar in molluscs by IC-ICP-MS and IC-ESI-MS/MS. J Anal At Spectrom 19:1454–1459.

Fujihara J, Kunito T, Kubota R, Tanabe S (2003) Arsenic accumulation in livers of pinnipeds, seabirds, and sea turtles: subcellular distribution and interaction between arsenobetaine and glycine betaine. Comp Biochem Physiol Part C 136:287–296.

Fujihara J, Kunito T, Kubota R, Tanaka H, Tanabe S (2004) Arsenic accumulation and distribution in tissues of black-footed albatrosses. Mar Pollut Bull 48:1153–1160.

Gailer J (2007) Arsenic-selenium and mercury-selenium bonds in biology. Coord Chem Rev 251:234–254.

Gailer J, Francesconi KA, Edmonds JS, Irgolic KJ (1995) Metabolism of arsenic compounds by the blue mussel *Mytilus edulis* after accumulation from seawater spiked with arsenic compounds. Appl Organomet Chem 9:341–355.

Gailer J, George GN, Pickering IJ, Prince RC, Ringwald SC, Pemberton JE, Glass RS, Younis HS, DeYoung DW, Aposhian HV (2000) A metabolic link between arsenite and selenite: the seleno-bis(*S*-glutathionyl) arsinium ion. J Am Chem Soc 122:4637–4639.

Gailer J, George GN, Pickering IJ, Prince RC, Younis HS, Winzerling JJ (2002a) Biliary excretion of $[(GS)_2AsSe]^-$ after intravenous injection of rabbits with arsenite and selenate. Chem Res Toxicol 15:1466–1471.

Gailer J, George GN, Harris HH, Pickering IJ, Prince RC, Somogyi A, Buttigieg GA, Glass RS, Denton MB (2002b) Synthesis, purification, and structural characterization of the dimethyldiselenoarsinate anion. Inorg Chem 41:5426–5432.

Geiszinger A, Goessler W, Kuehnelt D, Francesconi K, Kosmus W (1998) Determination of arsenic compounds in earthworms. Environ Sci Technol 32:2238–2243.

Geiszinger A, Khokiattiwong S, Goessler W, Francesconi KA (2002) Identification of the new arsenic-containing betaine, trimethylarsoniopropionate, in tissues of a stranded sperm whale *Physeter catodon*. J Mar Biol Assoc U K 82:165–168.

Gibbs PE, Langston WJ, Burt GR, Pascoe PL (1983) *Tharyx marioni* (Polychaeta): a remarkable accumulator of arsenic. J Mar Biol Assoc U K 63:313–325.

Goessler W, Rudorfer A, Mackey EA, Becker PR, Irgolic KJ (1998) Determination of arsenic compounds in marine mammals with high-performance liquid chromatography and an inductively coupled plasma mass spectrometer as element-specific detector. Appl Organomet Chem 12:491–501.

Gómes-Ariza JL, Sánchez-Rodas D, Giráldez I, Morales E (2000) A comparison between ICP-MS and AFS detection for arsenic speciation in environmental samples. Talanta 51:257–268.

Gorby MS (1994) Arsenic in human medicine. In: Nriagu JO (ed) Arsenic in the Environment, Part II: Human Health and Ecosystem Effects. Wiley, New York, pp 1–16.

Hanaoka K, Kaise T (1999) Microbial degradation of arsenobetaine accumulated in marine animals. J Natl Fish Univ 48:41–47.

Hanaoka K, Tagawa S, Kaise T (1992) The fate of organoarsenic compounds in marine ecosystems. Appl Organomet Chem 6:139–146.

Hanaoka K, Kaise T, Kai N, Kawasaki Y, Miyashita H, Kakimoto K, Tagawa S (1997) Arsenobetaine-decomposing ability of marine microorganisms occurring in particles collected at depths of 1100 and 3500 meters. Appl Organomet Chem 11:265–271.

Hanaoka K, Goessler W, Yoshida K, Fujitaka Y, Kaise T, Irgolic KJ (1999) Arsenocholine- and dimethylated arsenic-containing lipids in starspotted shark *Mustelus manazo*. Appl Organomet Chem 13:765–770.

Hanaoka K, Ohno H, Wada N, Ueno S, Goessler W, Kuehnelt D, Schlagenhaufen C, Kaise T, Irgolic KJ (2001a) Occurrence of organo-arsenicals in jellyfishes and their mucus. Chemosphere 44:743–749.

Hanaoka K, Tanaka Y, Nagata Y, Yoshida K, Kaise T (2001b) Water-soluble arsenic residues from several arsenolipids occurring in the tissues of the starspotted shark *Musterus manazo*. Appl Organomet Chem 15:299–305.

Hansen HR, Raab A, Francesconi KA, Feldmann J (2003) Metabolism of arsenic by sheep chronically exposed to arsenosugars as a normal part of their diet. 1. Quantitative intake, uptake, and excretion. Environ Sci Technol 37:845–851.

Hansen HR, Pickford R, Thomas-Oates J, Jaspars M, Feldmann J (2004a) 2-Dimethylarsinothioyl acetic acid identified in a biological sample: the first occurrence of a mammalian arsinothioyl metabolite. Angew Chem 116:341–344.

Hansen HR, Raab A, Jaspars M, Milne BF, Feldmann J (2004b) Sulfur-containing arsenical mistaken for dimethylarsinous acid [DMA(III)] and identified as a natural metabolite in urine: major implications for studies on arsenic metabolism and toxicity. Chem Res Toxicol 17:1086–1091.

Hansen HR, Jaspars M, Feldmann J (2004c) Arsinothioyl-sugars produced by *in vitro* incubation of seaweed extract with liver cytosol analysed by HPLC coupled simultaneously to ES-MS and ICP-MS. Analyst 129:1058–1064.

Haraguchi H (2004) Metallomics as integrated biometal science. J Anal At Spectrom 19:4–15.

Hasegawa H (1996) Seasonal changes in methylarsenic distribution in Tosa Bay and Uranouchi Inlet. Appl Organomet Chem 10:733–740.

Hayakawa T, Kobayashi Y, Cui X, Hirano S (2005) A new metabolic pathway of arsenite: arsenic-glutathione complexes are substrates for human arsenic methyltransferase Cyt19. Arch Toxicol 79:183–191.

Hei TK, Filipic M (2004) Role of oxidative damage in the genotoxicity of arsenic. Free Radic Biol Med 37:574–581.
Hirata S, Toshimitsu H, Aihara M (2006) Determination of arsenic species in marine samples by HPLC-ICP-MS. Anal Sci 22:39–43.
Ikemoto T, Kunito T, Tanaka H, Baba N, Miyazaki N, Tanabe S (2004) Detoxification mechanism of heavy metals in marine mammals and seabirds: interaction of selenium with mercury, silver, copper, zinc, and cadmium in liver. Arch Environ Contam Toxicol 47:402–413.
Jankong P, Chalhoub C, Kienzl N, Goessler W, Francesconi KA, Visoottiviseth P (2007) Arsenic accumulation and speciation in freshwater fish living in arsenic-contaminated waters. Environ Chem 4:11–17.
Jenkins RO, Ritchie AW, Edmonds JS, Goessler W, Molenat N, Kuehnelt D, Harrington CF, Sutton PG (2003) Bacterial degradation of arsenobetaine via dimethylarsinoylacetate. Arch Microbiol 180:142–150.
Kaise T, Hanaoka K, Tagawa S, Hirayama T, Fukui S (1988) Distribution of inorganic arsenic and methylated arsenic in marine organisms. Appl Organomet Chem 2:539–546.
Kanaki K, Pergantis SA (2007) HPLC-ICP-MS and HPLC-ES-MS/MS characterization of synthetic seleno-arsenic compounds. Anal Bioanal Chem 387:2617–2622.
Khokiattiwong S, Goessler W, Pedersen SN, Cox R, Francesconi KA (2001) Dimethylarsinoylacetate from microbial demethylation of arsenobetaine in seawater. Appl Organomet Chem 15:481–489.
Kirby J, Maher W (2002) Tissue accumulation and distribution of arsenic compounds in three marine fish species: relationship to trophic position. Appl Organomet Chem 16:108–115.
Kirby J, Maher W, Spooner D (2005) Arsenic occurrence and species in near-shore macroalgae-feeding marine animals. Environ Sci Technol 39:5999–6005.
Kitchin KT (2001) Recent advances in arsenic carcinogenesis: modes of action, animal model systems, and methylated arsenic metabolites. Toxicol Appl Pharmacol 172:249–261.
Kitchin KT, Ahmad S (2003) Oxidative stress as a possible mode of action for arsenic carcinogenesis. Toxicol Lett 137:3–13.
Koch I, Mace JV, Reimer KJ (2005) Arsenic speciation in terrestrial birds from Yellowknife Northwest Territories, Canada: the unexpected finding of arsenobetaine. Environ Toxicol Chem 24:1468–1474.
Kohlmeyer U, Jakubik S, Kuballa J, Jantzen E (2005) Determination of arsenic species in fish oil after acid digestion. Microchim Acta 151:249–255.
Kubota R, Kunito T, Tanabe S (2001) Arsenic accumulation in the liver tissue of marine mammals. Environ Pollut 115:303–312.
Kubota R, Kunito T, Tanabe S (2002a) Chemical speciation of arsenic in the livers of higher trophic marine animals. Mar Pollut Bull 45:218–223.
Kubota R, Kunito T, Tanabe S, Ogi H, Shibata Y (2002b) Maternal transfer of arsenic to eggs of black-tailed gull (*Larus crassirostis*) from Rishiri Island, Japan. Appl Organomet Chem 16:463–468.
Kubota R, Kunito T, Tanabe S (2003a) Occurrence of several arsenic compounds in the liver of birds, cetaceans, pinnipeds, and sea turtles. Environ Toxicol Chem 22:1200–1207.
Kubota R, Kunito T, Tanabe S (2003b) Is arsenobetaine the major arsenic compound in the liver of birds, marine mammals, and sea turtles? J Phys IV 107:707–710.
Kubota R, Kunito T, Fujihara J, Tanabe S, Yang J, Miyazaki N (2005) Placental transfer of arsenic to fetus of Dall's porpoises (*Phocoenoides dalli*). Mar Pollut Bull 51:845–849.
Kubota R, Kunito T, Agusa T, Fujihara J, Monirith I, Iwata H, Subramanian A, Tana TS, Tanabe S (2006) Urinary 8-hydroxy-2'-deoxyguanosine in inhabitants chronically exposed to arsenic in groundwater in Cambodia. J Environ Monit 8:293–299.
Kuehnelt D, Goessler W (2003) Organoarsenic compounds in the terrestrial environment. In: Craig PJ (ed) Organometallic Compounds in the Environment. Wiley, New York, pp 223–275.
Kuehnelt D, Goessler W, Irgolic KJ (1997a) Arsenic compounds in terrestrial organisms I: *Collybia maculata*, *Collybia butyracea* and *Amanita muscaria* from arsenic smelter sites in Austria. Appl Organomet Chem 11:289–296.
Kuehnelt D, Goessler W, Schlagenhaufen C, Irgolic KJ (1997b) Arsenic compounds in terrestrial organisms III: arsenic compounds in *Formica* sp. from an old arsenic smelter site. Appl Organomet Chem 11:859–867.

Lam JCW, Tanabe S, Chan SKF, Lam MHW, Martin M, Lam PKS (2006) Levels of trace elements in green turtle eggs collected from Hong Kong: evidence of risks due to selenium and nickel. Environ Pollut 144:790–801.

Langlois C, Langis R (1995) Presence of airborne contaminants in the wildlife of northern Quebec. Sci Total Environ 160/161:391–402.

Langston WJ, Spence SK (1995) Biological factors involved in metal concentrations observed in aquatic organisms. In: Tessier A, Turner DR (eds) Metal Speciation and Bioavailability in Aquatic Systems. Wiley, Chichester, UK, pp 407–478.

Larsen EH, Francesconi KA (2003) Arsenic concentrations correlate with salinity for fish taken from the North Sea and Baltic waters. J Mar Biol Assoc U K 83:283–284.

Law RJ (1996) Metals in marine mammals. In: Beyer WN, Heinz GH, Redmon-Norwood AW (eds) Environmental Contaminants in Wildlife: Interpreting Tissue Concentrations. CRC Press, Boca Raton, pp 357–376.

Le XC, Lu X, Li X-F (2004) Arsenic speciation. Anal Chem 76:27A–33A.

Lien YHH, Pacelli MM, Braun EJ (1993) Characterization of organic osmolytes in avian renal medulla: a nonurea osmotic gradient system. Am J Physiol 264:R1045–R1049.

Lunde G (1977) Occurrence and transformation of arsenic in the marine environment. Environ Health Perspect 19:47–52.

Mancini I, Guella G, Frostin M, Hnawia E, Laurent D, Debitus C, Pietra F (2006) On the first polyarsenic organic compound from nature: arsenicin A from the New Caledonian marine sponge *Echinochalina bargibanti*. Chem Eur J 12:8989–8994.

Mandal BK, Suzuki KT (2002) Arsenic round the world: a review. Talanta 58:201–235.

Manley SA, George GN, Pickering IJ, Glass RS, Prenner EJ, Yamdagni R, Wu Q, Gailer J (2006) The seleno bis(*S*-glutathionyl) arsinium ion is assembled in erythrocyte lysate. Chem Res Toxicol 19:601–607.

Martin SJ, Newcombe C, Raab A, Feldmann J (2005) Arsenosugar metabolism not unique to the sheep of North Ronaldsay. Environ Chem 2:190–197.

McSheehy S, Szpunar J, Lobinski R, Haldys V, Tortajada J, Edmonds JS (2002) Characterization of arsenic species in kidney of the clam *Tridacna derasa* by multidimensional liquid chromatography-ICPMS and electrospray time-of-flight tandem mass spectrometry. Anal Chem 74:2370–2378.

Meador JP, Varanasi U, Robisch PA, Chan SL (1993) Toxic metals in pilot whales (*Globicephala melaena*) from strandings in 1986 and 1990 on Cape Cod, Massachusetts. Can J Fish Aquat Sci 50:2698–2706.

Meier J, Kienzl N, Goessler W, Francesconi KA (2005) The occurrence of thio-arsenosugars in some samples of marine algae. Environ Chem 2:304–307.

Miyajima M, Hamada N, Yoshimura E, Okubo A, Yamazaki S, Toda S (1988) Lipophilic arsenic compound(s) in the liver of a tiger shark (*Galeocerdo cuvier*). Appl Organomet Chem 2:377–384.

Morita M, Shibata Y (1988) Isolation and identification of arseno-lipid from a brown alga, *Undaria pinnatifida* (Wakame). Chemosphere 17:1147–1152.

Morita M, Shibata Y (1990) Chemical form of arsenic in marine macroalgae. Appl Organomet Chem 4:181–190.

Morton WE, Dunnette DA (1994) Health effects of environmental arsenic. In: Nriagu JO (ed) Arsenic in the Environment, Part II: Human Health and Ecosystem Effects. Wiley, New York, pp 17–34.

Nakazato T, Tao H (2006) A high-efficiency photooxidation reactor for speciation of organic arsenicals by liquid chromatography-hydride generation-ICPMS. Anal Chem 78:1665–1672.

Naranmandura H, Suzuki N, Suzuki KT (2006) Trivalent arsenicals are bound to proteins during reductive methylation. Chem Res Toxicol 19:1010–1018.

Neff JM (1997) Ecotoxicology of arsenic in the marine environment. Environ Toxicol Chem 16:917–927.

Ng PS, Li H, Matsumoto K, Yamazaki S, Kogure T, Tagai T, Nagasawa H (2001) Striped dolphin detoxificates mercury as insoluble Hg(S, Se) in the liver. Proc Jpn Acad 77(Ser B):178–183.

Ninh TD, Nagashima Y, Shiomi K (2007) Water-soluble and lipid-soluble arsenic compounds in Japanese flying squid *Todarodes pacificus*. J Agric Food Chem 55:3196–3202.

Nischwitz V, Pergantis SA (2005a) First report on the detection and quantification of arsenobetaine in extracts of marine algae using HPLC-ES-MS/MS. Analyst 130:1348–1350.
Nischwitz V, Pergantis SA (2005b) Liquid chromatography online with selected reaction monitoring electrospray mass spectrometry for the determination of organoarsenic species in crude extracts of marine reference materials. Anal Chem 77:5551–5563.
Nischwitz V, Pergantis SA (2006) Optimisation of an HPLC selected reaction monitoring electrospray tandem mass spectrometry method for the detection of 50 arsenic species. J Anal At Spectrom 21:1277–1286.
Nischwitz V, Kanaki K, Pergantis SA (2006) Mass spectrometric identification of novel arsinothioyl-sugars in marine bivalves and algae. J Anal At Spectrom 21:33–40.
Nordstrom DK (2002) Worldwide occurrences of arsenic in ground water. Science 296:2143–2145.
Nriagu JO (1989) A global assessment of natural sources of atmospheric trace metals. Nature (Lond) 338:47–49.
Nriagu JO (2002) Arsenic poisoning through the ages. In: Frankenberger WT Jr (ed) Environmental Chemistry of Arsenic. Dekker, New York, pp 1–26.
Oremland RS, Stolz JF (2003) The ecology of arsenic. Science 300:939–944.
O'Shea TJ (1999) Environmental contaminants and marine mammals. In: Reynolds JE III, Rommel SA (eds) Biology of Marine Mammals. Smithsonian Institution Press, Washington, pp 485–563.
O'Shea TJ, Tanabe S (2003) Persistent ocean contaminants and marine mammals: a retrospective overview. In: Vos JG, Bossart GD, Fournier M, O'Shea TJ (eds) Toxicology of Marine Mammals. Taylor & Francis, London, pp 99–134.
Pacyna JM, Pacyna EG (2001) An assessment of global and regional emissions of trace metals to the atmosphere from anthropogenic sources worldwide. Environ Rev 9:269–298.
Phillips DJH (1990) Arsenic in aquatic organisms: a review, emphasizing chemical speciation. Aquat Toxicol 16:151–186.
Pichereau V, Cosquer A, Gaumont AC, Bernard T (1997) Synthesis of trimethylated phosphonium and arsonium analogues of the osmoprotectant glycine betaine; contrasted biological activities in two bacterial species. Bioorg Med Chem Lett 7:2893–2896.
Plant JA, Kinniburgh DG, Smedley PL, Fordyce FM, Klinck BA (2005) Arsenic and selenium. In: Lollar BS (ed) Environmental Geochemistry. Elsevier, Amsterdam, pp 17–66.
Raab A, Wright SH, Jaspars M, Meharg AA, Feldmann J (2007) Pentavalent arsenic can bind to biomolecules. Angew Chem Int Ed 46:2594–2597.
Raml R, Goessler W, Traar P, Ochi T, Francesconi KA (2005) Novel thioarsenic metabolites in human urine after ingestion of an arsenosugar, 2',3'-dihydroxypropyl 5-deoxy-5-dimethylarsinoyl-β-D-riboside. Chem Res Toxicol 18:1444–1450.
Raml R, Goessler W, Francesconi KA (2006) Improved chromatographic separation of thioarsenic compounds by reversed-phase high performance liquid chromatography-inductively coupled plasma mass spectrometry. J Chromatogr A 1128:164–170.
Raml R, Rumpler A, Goessler W, Vahter M, Li L, Ochi T, Francesconi KA (2007) Thio-dimethylarsinate is a common metabolite in urine samples from arsenic-exposed women in Bangladesh. Toxicol Appl Pharmacol 222:374–380.
Randall K, Lever M, Peddie BA, Chambers ST (1995) Competitive accumulation of betaines by *Escherichia coli* K-12 and derivative strains lacking betaine porters. Biochim Biophys Acta 1245:116–120.
Randall K, Lever M, Peddie BA, Chambers ST (1996) Accumulation of natural and synthetic betaines by a mammalian renal cell line. Biochem Cell Biol 74:283–287.
Ritchie AW, Edmonds JS, Goessler W, Jenkins RO (2004) An origin for arsenobetaine involving bacterial formation of an arsenic-carbon bond. FEMS Microbiol Lett 235:95–99.
Saeki K, Sakakibara H, Sakai H, Kunito T, Tanabe S (2000) Arsenic accumulation in three species of sea turtles. BioMetals 13:241–250.
Santosa SJ, Mokudai H, Takahashi M, Tanaka S (1996) The distribution of arsenic compounds in the ocean: biological activity in the surface zone and removal processes in the deep zone. Appl Organomet Chem 10:697–705.

Schaeffer R, Francesconi KA, Kienzl N, Soeroes C, Fodor P, Váradi L, Raml R, Goessler W, Kuehnelt D (2006) Arsenic speciation in freshwater organisms from the river Danube in Hungary. Talanta 69:856–865.

Schmeisser E, Raml R, Francesconi KA, Kuehnelt D, Lindberg AL, Sörös C, Goessler W (2004) Thio arsenosugars identified as natural constituents of mussels by liquid chromatography-mass spectrometry. Chem Commun 2004:1824–1825

Schmeisser E, Goessler W, Kienzl N, Francesconi KA (2005) Direct measurement of lipid-soluble arsenic species in biological samples with HPLC-ICPMS. Analyst 130:948–955.

Schmeisser E, Rumpler A, Kollroser M, Rechberger G, Goessler W, Francesconi KA (2006a) Arsenic fatty acids are human urinary metabolites of arsenolipids present in cod liver. Angew Chem Int Ed 45:150–154.

Schmeisser E, Goessler W, Francesconi KA (2006b) Human metabolism of arsenolipids present in cod liver. Anal Bioanal Chem 385:367–376.

Schwerdtle T, Walter I, Mackiw I, Hartwig A (2003) Induction of oxidative DNA damage by arsenite and its trivalent and pentavalent methylated metabolites in cultured human cells and isolated DNA. Carcinogenesis 24:967–974.

Shaw JR, Gabor K, Hand E, Lankowski A, Durant L, Thibodeau R, Stanton CR, Barnaby R, Coutermarsh B, Karlson KH, Sato JD, Hamilton JW, Stanton BA (2007) Role of glucocorticoid receptor in acclimation of killifish (*Fundulus heteroclitus*) to seawater and effects of arsenic. Am J Physiol Regul Integr Comp Physiol 292:R1052–R1060.

Shibata Y, Morita M (1992) Characterization of organic arsenic compounds in bivalves. Appl Organomet Chem 6:343–349.

Shibata Y, Morita M (2000) Chemical forms of arsenic in the environment. Biomed Res Trace Elements 11:1–24 (in Japanese).

Shibata Y, Morita M, Fuwa K (1992) Selenium and arsenic in biology: their chemical forms and biological functions. Adv Biophys 28:31–80.

Shinagawa A, Shiomi K, Yamanaka H, Kikuchi T (1983) Selective determination of inorganic arsenic (III), (V) and organic arsenic in marine organisms. Bull Jpn Soc Sci Fish 49:75–78.

Shiomi K (1994) Arsenic in marine organisms: chemical forms and toxicological aspects. In: Nriagu JO (ed) Arsenic in the Environment, Part II: Human Health and Ecosystem Effects. Wiley, New York, pp 261–282.

Shiomi K, Sugiyama Y, Shimakura K, Nagashima Y (1996) Retention and biotransformation of arsenic compounds administered intraperitoneally to carp. Fish Sci 62:261–266.

Šlejkovec Z, Bajc Z, Doganoc DZ (2004) Arsenic speciation patterns in freshwater fish. Talanta 62:931–936.

Sloth JJ, Larsen EH, Julshamn K (2003) Determination of organoarsenic species in marine samples using gradient elution cation exchange HPLC-ICP-MS. J Anal At Spectrom 18:452–459.

Sloth JJ, Larsen EH, Julshamn K (2005a) Report on three aliphatic dimethylarsinoyl compounds as common minor constituents in marine samples. An investigation using high-performance liquid chromatography/inductively coupled plasma mass spectrometry and electrospray ionisation tandem mass spectrometry. Rapid Commun Mass Spectrom 19:227–235.

Sloth JJ, Larsen EH, Julshamn K (2005b) Survey of inorganic arsenic in marine animals and marine certified reference materials by anion exchange high-performance liquid chromatography–inductively coupled plasma mass spectrometry. J Agric Food Chem 53:6011–6018.

Soeroes C, Goessler W, Francesconi KA, Kienzl N, Schaeffer R, Fodor P, Kuehnelt D (2005) Arsenic speciation in farmed Hungarian freshwater fish. J Agric Food Chem 53:9238–9243.

Stanton CR, Thibodeau R, Lankowski A, Shaw JR, Hamilton JW, Stanton BA (2006) Arsenic inhibits CFTR-mediated chloride secretion by killifish (*Fundulus heteroclitus*) opercular membrane. Cell Physiol Biochem 17:269–278.

Stoica A, Pentecost E, Martin MB (2000) Effects of arsenite on estrogen receptor-α expression and activity in MCF-7 breast cancer cells. Endocrinology 141:3595–3602.

Storelli MM, Marcotrigiano GO (2000) Total organic and inorganic arsenic from marine turtles (*Caretta caretta*) beached along the Italian coast (South Adriatic Sea). Bull Environ Contam Toxicol 65:732–739.

Styblo M, Thomas DJ (1997) Binding of arsenicals to proteins in an *in vitro* methylation system. Toxicol Appl Pharmacol 147:1–8.

Styblo M, Delnomdedieu M, Thomas DJ (1996) Mono- and dimethylation of arsenic in rat liver cytosol *in vitro*. Chem-Biol Interact 99:147–164.

Suedel BC, Boraczek JA, Peddicord RK, Clifford PA, Dillon TM (1994) Trophic transfer and biomagnification potential of contaminants in aquatic ecosystems. Rev Environ Contam Toxicol 136:21–89.

Suzuki KT (2005) Metabolomics of arsenic based on speciation studies. Anal Chim Acta 540:71–76.

Tanabe S, Subramanian A (2006) Bioindicators of POPs: Monitoring in Developing Countries. Kyoto University Press, Kyoto, Japan, pp 190.

Tanabe S, Tatsukawa R, Maruyama K, Miyazaki N (1982) Transplacental transfer of PCBs and chlorinated hydrocarbon pesticides from the pregnant striped dolphin (*Stenella coeruleoalba*) to her fetus. Agric Biol Chem 46:1249–1254.

Thompson DR (1990) Metal levels in marine vertebrates. In: Furness RW, Rainbow PS (eds) Heavy Metals in the Marine Environment. CRC Press, Boca Raton, pp 143–182.

Vahter M (1999) Methylation of inorganic arsenic in different mammalian species and population groups. Sci Prog 82:69–88.

Vahter M, Marafante E, Dencker L (1983) Metabolism of arsenobetaine in mice, rats and rabbits. Sci Total Environ 30:197–211.

Waalkes MP, Liu J, Chen H, Xie Y, Achanzar WE, Zhou Y-S, Cheng M-L, Diwan BA (2004) Estrogen signaling in livers of male mice with hepatocellular carcinoma induced by exposure to arsenic *in utero*. J Natl Cancer Inst 96:466–474.

Wahlen R, McSheehy S, Scriver C, Mester Z (2004) Arsenic speciation in marine certified reference materials. Part 2. The quantification of water-soluble arsenic species by high-performance liquid chromatography-inductively coupled plasma mass spectrometry. J Anal At Spectrom 19:876–882.

Watanabe I, Kunito T, Tanabe S, Amano M, Koyama Y, Miyazaki N, Petrov EA, Tatsukawa R (2002) Accumulation of heavy metals in Caspian seals (*Phoca caspica*). Arch Environ Contam Toxicol 43:109–120.

WHO (2001) Environmental Health Criteria 224: Arsenic and Arsenic Compounds, 2nd Ed. World Health Organization, Geneva.

Yamanaka K, Kato K, Mizoi M, An Y, Takabayashi F, Nakano M, Hoshino M, Okada S (2004) The role of active arsenic species produced by metabolic reduction of dimethylarsinic acid in genotoxicity and tumorigenesis. Toxicol Appl Pharmacol 198:385–393.

Yancey PH, Clark ME, Hand SC, Bowlus RD, Somero GN (1982) Living with water stress: evolution of osmolyte systems. Science 217:1214–1222.

Yoshida K, Kuroda K, Inoue Y, Chen H, Wanibuchi H, Fukushima S, Endo G (2001) Metabolites of arsenobetaine in rats: does decomposition of arsenobetaine occur in mammals? Appl Organomet Chem 15:271–276.

Yoshitome R, Kunito T, Ikemoto T, Tanabe S, Zenke H, Yamauchi M, Miyazaki N (2003) Global distribution of radionuclides (^{137}Cs and ^{40}K) in marine mammals. Environ Sci Technol 37:4597–4602.

Zhu J, Chen Z, Lallemand-Breitenbach V, de Thé H (2002) How acute promyelocytic leukaemia revived arsenic. Nat Rev Cancer 2:705–713.

Environmental Chemistry, Ecotoxicity, and Fate of Lambda-Cyhalothrin

Li-Ming He, John Troiano, Albert Wang, and Kean Goh

L.-M. He (✉)
Surface Water Protection Program, Environmental Monitoring Branch, Department of Pesticide Regulation, California Environmental Protection Agency, 1001 I Street, Sacramento, CA 95814, USA (e-mail: lhe@cdpr.ca.gov)

J. Troiano
Surface Water Protection Program, Environmental Monitoring Branch, Department of Pesticide Regulation, California Environmental Protection Agency, 1001 I Street, Sacramento, CA 95814, USA

K.S. Goh
Surface Water Protection Program, Environmental Monitoring Branch, Department of Pesticide Regulation, California Environmental Protection Agency, 1001 I Street, Sacramento, CA 95814, USA

A. Wang
Office of Environmental Health Hazard Assessment, California Environmental Protection Agency, P.O. Box 4010, Sacramento, CA 95812-4010, USA

D.M. Whitacre (ed.), *Reviews of Environmental Contamination and Toxicology.*

1 Introduction

Lambda-cyhalothrin is a pyrethroid insecticide. Pyrethroids are synthetic chemical analogues of pyrethrins, which are naturally occurring insecticidal compounds produced in the flowers of chrysanthemums (*Chrysanthemum cinerariaefolium*). Insecticidal products containing pyrethroids have been widely used to control insect pests in agriculture, public health, and homes and gardens (Amweg and Weston 2005; Oros and Werner 2005). In agriculture, target crops include cotton, cereals, hops, ornamentals, potatoes, and vegetables, with applications made to control aphid, coleopterous, and lepidopterous pests. Pyrethroids are important tools used in public health management where applications are made to control cockroaches, mosquitoes, ticks, and flies, which may act as disease vectors. Residential use of pyrethroid products has increased because of the suspension of organophosphate products containing chlorpyrifos or diazinon (Oros and Werner 2005; Weston et al. 2005).

Lambda-cyhalothrin is the active ingredient (a.i.) in several brand name products: Warrior, Scimitar, Karate, Demand, Icon, and Matador. Annual agricultural use of lambda-cyhalothrin in California has been consistent at approximately 30,000 lbs a.i. per annum from 2000 to 2003 and increased to ~40,000 lbs a.i. per annum between 2004 and 2006 (CDPR 2006). Residues of lambda-cyhalothrin have been detected in irrigation and storm runoff water and in their associated sediments. Residues have been detected in runoff resulting from agricultural, public health, and residential applications. For example, lambda-cyhalothrin was detected in water at 0.11–0.14 μg/L from agricultural watersheds in Stanislaus County, California. Lambda-cyhalothrin residues were detected in sediments obtained from sites sampled in Imperial, Monterey, Stanislaus, and Placer Counties. Residues in sediment ranged from 0.003 to 0.315 μg/g of dry wt (Starner 2007).

Toxicity tests conducted at levels of lambda-cyhalothrin residues measured in water or sediment indicate potential for effects on aquatic organisms including fish and amphipods (Amweg et al. 2005, 2006; Cavas and Ergene-Gozukara 2003; Gu et al. 2007; Heckmann and Friberg 2005; Lawler et al. 2007; Maund et al. 1998; Van Wijngaarden et al. 2005; Wang et al. 2007; Weston et al. 2004). Concerns have therefore been raised about the widespread use of lambda-cyhalothrin in California and its potential impact on aquatic ecosystems. This review is limited to literature available from peer-reviewed publications and approved documents and databases and is not deemed to be exhaustive. Research results from studies using multiple pyrethroids that included lambda-cyhalothrin were used in this review to augment data from studies conducted solely with lambda-cyhalothrin.

2 Chemistry of Lambda-Cyhalothrin

Lambda-cyhalothrin is a 1:1 mixture of two stereoisomers, (*S*)-α-cyano-3-phenoxybenzyl-(*Z*)-(1*R*,3*R*)-3-(2-chloro-3,3,3-trifluoroprop-1-enyl)-2,2-dimethyl cyclopropanecarboxylate (Fig. 1a) and (*R*)-α-cyano-3-phenoxybenzyl-(*Z*) -(1*S*,3*S*)-

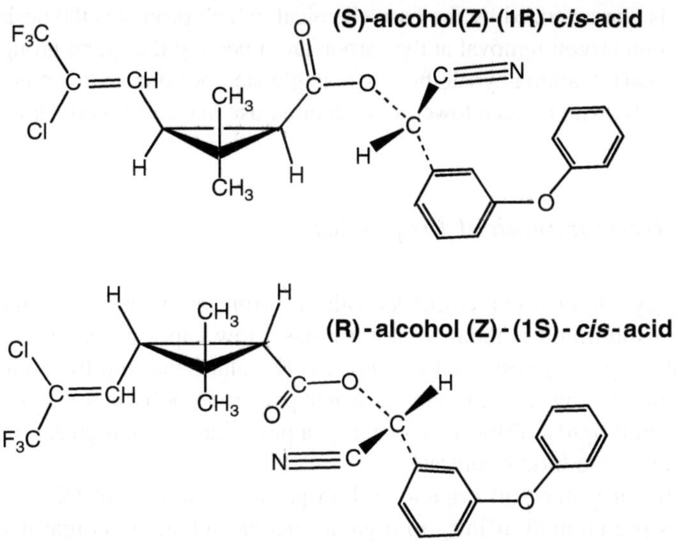

a

b

Fig. 1 The chemical structure of two isomers of lambda-cyhalothrin

3-(2-chloro-3, 3,3-trifluoroprop-1-enyl)-2,2-dimethylcyclopropanecarboxylate (Fig. 1b). Lambda-cyhalothrin was first reported by Robson and Crosby (1984) and was introduced in Central America and the Far East in 1985 by ICI Agrochemicals (now Syngenta).

2.1 *Synthesis*

Dried and ground pyrethrum chrysanthemum flowers were noted to be powerful insecticides in ancient China, but it took until the middle of the 20th century for the first improved chemical analogues of the natural pyrethrin a.i.s to be commercially synthesized. Naturally occurring pyrethrins are esters consisting of a so-called "acid" component, which has a cyclopropane core and an "alcohol" component. At first, synthetic pyrethroids were developed through alcohol substitutions. This early generation of synthetic pyrethroids had drawbacks, mostly the result of poor stability in sunlight. Photostability was greatly improved by substituting vinylic halogens in the acid component (Spurlock 2006). Additional α-cyano-substitution in the alcohol component by the National Research Development Corporation in the UK further improved light stability and insecticidal activity. The commercial product KARATE, whose a.i. is lambda-cyhalothrin, was registered for use by the U.S. Environmental Protection Agency (USEPA) in 1988 (Syngenta 2007).

It is known that some isomers have a greater insecticidal effect than others, and this has led to the development of techniques to separate the more-active isomers and to convert the less-active isomers into the more-active ones. Conversion between isomers is characterized by the base chemical, which promotes the desired epimerization through proton removal at the carbon atom bearing the cyano group (Cleugh and Milner 1994). Gamma-cyhalothrin is a single stereoisomer that has effective insecticidal activity, with a much lower total reported use in California (CDPR 2007).

2.2 *Physicochemical Properties*

Lambda-cyhalothrin is a colorless solid at room temperature but may appear yellowish in solution. Lambda-cyhalothrin has a low vapor pressure and Henry's law constant, which suggests that it is not easily volatilized into the atmosphere. This insecticide also has a high octanol–water partition coefficient (K_{ow}), so it tends to partition into lipids (Table 1). Normally, a pesticide with a high K_{ow} would signal a high potential to bioconcentrate.

The mean water–soil organic carbon partition coefficient (K_{oc}) is high, which indicates preferential affinity to organic matter and suggests that it is unlikely to contaminate groundwater because of a low potential to leach as dissolved residues in percolating water. The tendency to adsorb to suspended particulate materials in

Table 1 Physical, chemical, and environmental properties of lambda-cyhalothrin

CAS number	91465-08-6
US EPA PC Code	128897
CA DPR Chem Code	2297
Molecular formula	$C_{23}H_{19}ClF_3NO_3$
Molecular weight (g/mol)	449.9
Density (g/mL at 25°C)	1.33
Melting point (°C)	49.2
Boiling point (°C at 0.2 mmHg)	187–190
Vapor pressure (mPa at 20°C)	0.0002
Henry's law constant (Pa-m^3/mole)	0.018
Water solubility (mg/L at 20°C)	0.005
Solubility in other solvents (e.g., acetone) (mg/L)	>500,000
Octanol–water partitioning (log K_{ow} at 20°C)	7.00
Hydrolysis half-life (d)	
pH 5	Stable
pH 7	Stable
pH 9	8.66
Photolysis half-life (d)	
Water at pH 5 and 25°C	24.5
Soil	53.7
Bioconcentration factor (BCF) (fish)	2,240
Soil adsorption K_{oc} (cm^3/g)	247,000–330,000
Soil degradation half-life (d)	
Aerobic soil	42.6
Aquatic degradation half-life (d)	
aerobic aquatic	21.9

Source: CDPR 2007; Laskowski 2002; PAN 2007; Tomlin 2000; USDA 2007; USEPA 2007.

the water column, including clay particles and organic matter, provides the primary vector for transport through aquatic systems. Thus, the greatest risk to nontarget aquatic organisms would be through exposure to lambda-cyhalothrin-contaminated sediments. Adsorbed phases of chemical molecules generally show decreased degradation rates because residues are less accessible to breakdown by sunlight or microorganisms than when molecules are dissolved in the water column (Schwarzenbach et al. 1993). Sorption of lambda-cyhalothrin to suspended solids or bottom sediments may provide a mechanism to mitigate its acute toxicity to aquatic organisms by reducing its short-term bioavailability in the water column.

2.3 Mode of Action

Pyrethroids are axonic poisons that affect the nerve fiber by binding to a protein that regulates the voltage-gated sodium channel. Normally, this gate opens to cause stimulation of the nerve and closes to terminate the nerve signal. The channels are pathways through which ions are permitted to enter the axon and cause excitation. When the channels are left open, nerve cells produce repetitive discharges and eventually cause paralysis (Bradbury and Coats 1989; Shafer and Meyer 2004). Pyrethroids bind to this gate and prevent it from closing normally, which results in continuous nerve stimulation and tremors in poisoned insects. Poisoned organisms lose control of their nervous system and are unable to produce coordinated movement.

There are two groups of pyrethroids with distinctive poisoning symptoms, denoted as Type I and Type II. Chemically, Type II pyrethroids are distinguished from Type I pyrethroids by the presence of an α-cyano group in their structure. In comparison to Type I pyrethroids (e.g., permethrin), which exert their neurotoxicity primarily through interference with sodium channel function in the central nervous system, Type II pyrethroids (e.g., lambda-cyhalothrin) can also affect chloride and calcium channels that are important for proper nerve function (Burr and Ray 2004).

Because of the lipophilic nature of pyrethroids, biological membranes and tissues readily absorb them. Specifically, lambda-cyhalothrin penetrates the insect cuticle, disrupting nerve conduction within minutes; this leads to cessation of feeding, loss of muscular control, paralysis, and eventual death. Additional protection of the crop is provided by the insecticide's strong repellent effect toward insects.

3 Breakdown Mechanisms and Products

3.1 Photolysis

As indicated previously, the naturally occurring pyrethrins are unstable in light, while the photostability of recent synthetic pyrethroids has been improved. Photochemical studies with lambda-cyhalothrin were conducted under UV and sunlight irradiation to understand photodegradation kinetics, pathways, and products (Fernandez-Alvarez et al. 2007; Ruzo et al. 1987). Exposure to UV light (18 W, 254 nm) for 20 min resulted in nearly complete degradation with losses greater than 95% of initial amounts applied (Fernandez-Alvarez et al. 2007). Photodegradation of lambda-cyhalothrin followed first-order kinetic behavior where the apparent first-order rate constant (k_{ap}) and half-life ($t_{1/2}$) were determined to be 0.163 min^{-1} and 4.26 min, respectively.

With the recent development of highly efficient extraction methods and high-resolution detection techniques, it is possible to identify multiple trace photoproducts and photodegradation pathways for lambda-cyhalothrin. Several pathways have been proposed for the photodegradation of lambda-cyhalothrin, including decarboxylation, reductive dehalogenation, and ester or other bond cleavage (Fernandez-Alvarez et al. 2007) (Fig. 2).

The photoproduct decarboxycyhalothrin (P4) of lambda-cyhalothrin is generated by the decarboxylation pathway. The ester bond cleavage of lambda-

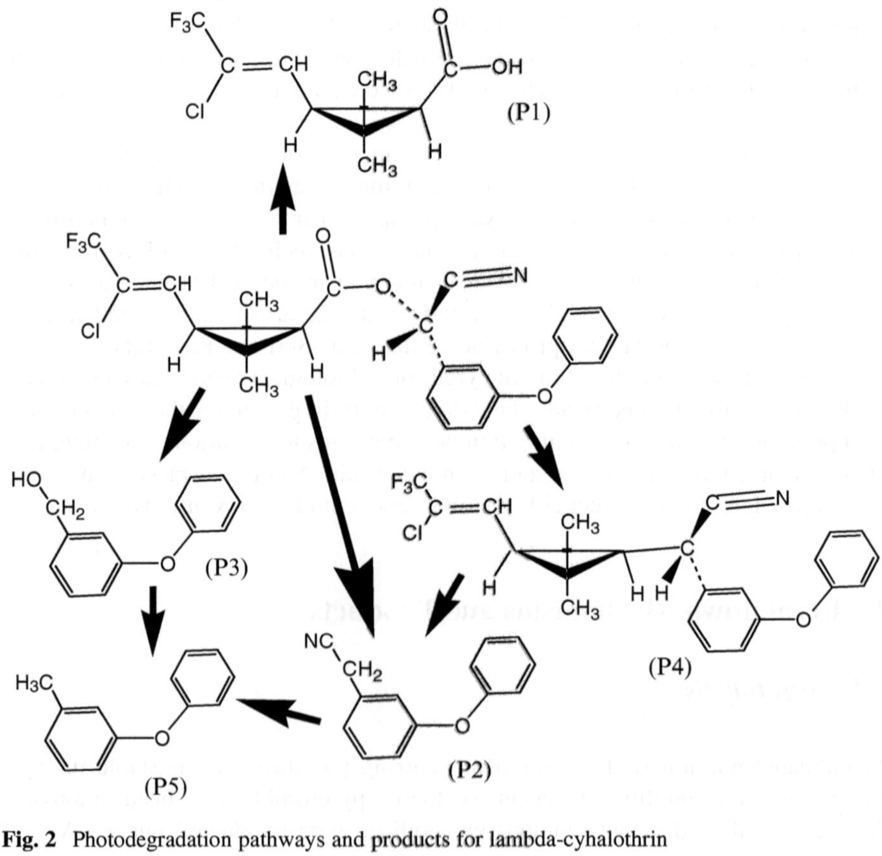

Fig. 2 Photodegradation pathways and products for lambda-cyhalothrin

cyhalothrin generates 3-(2-chloro-3,3,3-trifluoroprop-1-en-1-yl)-2,2-dimethyl cyclopropanecarboxylic acid (P1) and (3-phenoxyphenyl)acetonitrile (P2). The photoproduct P2 can also be generated by C–C bond cleavage of photoproduct P4. Photoproduct P3 ((3-phenoxyphenyl)methanol) can be generated by the ester cleavage and subsequent loss of the cyano group of lambda-cyhalothrin. The photoproduct 1-methyl-3-phenoxybenzene (P5) may be formed by hydroxyl or cyano group losses from P3 or P2 (Fernandez-Alvarez et al. 2007).

3.2 Hydrolysis

Lambda-cyhalothrin is stable at pH below 8, whereas under alkaline conditions it hydrolyzes through nucleophilic attack of the hydroxyl ion. A cyanohydrin derivative is formed, which degrades to yield HCN and the corresponding aldehyde (Fig. 3) (Gupta et al. 1998).

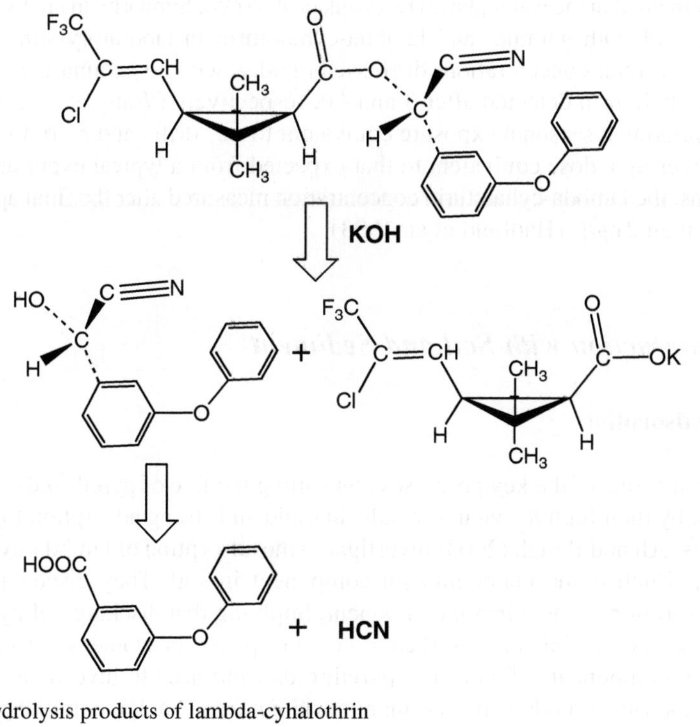

Fig. 3 Hydrolysis products of lambda-cyhalothrin

3.3 Microbial Degradation

In laboratory studies, the dissipation of lambda-cyhalothrin in soil was mainly through biodegradation, as indicated by the rapid loss of lambda-cyhalothrin in non-sterile soil compared to sterile soil (Wang et al. 1997). Only one degradate, which was (*RS*)-α-cyano-3-(4-hydroxyphenoxy)benzyl-(*Z*)-(1*RS*)-*cis*-3-(2-chloro-3, 3, 3-trifluoropropenyl)-2,2-dimethylcyclopropanecarboxylate, was determined as a major breakdown product comprising 10% of the initial lambda-cyhalothrin concentration (European-Commission 2001).

4 Environmental Fate of Lambda-Cyhalothrin

4.1 Dissipation in Water

Lambda-cyhalothrin rapidly dissipates from water (Farmer et al. 1995). The fate of lambda-cyhalothrin was compared in mesotrophic and eutrophic ditch microcosms where it was applied three times at 1-wk intervals at concentrations of 10, 25, 50, 100, and 250 ng/L. The rate of dissipation of lambda-cyhalothrin in the water column of the two systems was similar where, after 1 d, only 30% of the amount applied remained in the water phase (Roessink et al. 2005). Similarly, in an investigation of the fate of both gamma- and lambda-cyhalothrin in laboratory-simulated rice paddy water, their concentrations decreased rapidly, with no gamma-cyhalothrin or lambda-cyhalothrin detected after 3 and 4 d, respectively (Wang et al. 2007). Last, after simulating a seasonal exposure equivalent to 12 "drift" and 6 "runoff" events, each delivering a dose equivalent to that expected from a typical event under field conditions, the lambda-cyhalothrin concentration measured after the final application was less than 2 ng/L (Hadfield et al. 1993).

4.2 Interaction with Soil and Sediment

4.2.1 Adsorption

Adsorption is one of the key processes controlling the fate of pyrethroids, which, as indicated by their high K_{oc} values, results in rapid and strong adsorption to soils and sediments. Ali and Baugh (2003) investigated the adsorption of lambda-cyhalothrin to silica, which is the major mineral component in soil. They discovered much-reduced sorption to the mineral component, implying that discharge of pyrethroids to water bodies is mainly in the sorbed soil organic phase. In another study of sorption to mineral components of soil, four pyrethroids were used to investigate selective interactions with corundum, quartz, montmorillonite, and kaolinite. Initial pyrethroid concentrations ranged from 1 to 100 μg/L. Sorption to glass centrifuge tubes used in the batch experiments was significant and accounted for 25%–60% of total sorption (Oudou and Hansen 2002; Zhou et al. 1995). All corrected adsorption isotherms fit the Freundlich equation with exponential n values ranging between 0.9 and 1.1. These n values indicated little curvature in the adsorption isotherm. Bonding affinities per unit surface area decreased in the order corundum > quartz > montmorillonite > kaolinite. All minerals showed the same selectivity order with respect to sorption affinity of the four pyrethroids: lambda-cyhalothrin > deltamethrin > cypermethrin > fenvalerate (Oudou and Hansen 2002).

Adsorption is a surface phenomenon that for pyrethroids has been shown to depend on the surface area and the organic carbon content of the adsorption. Sorption–desorption equilibria of six pyrethroids (permethrin, cyfluthrin, cypermethrin, lambda-cyhalothrin, deltamethrin, and fenvalerate) were determined after 24-hr equilibrium in soils with an organic carbon content that ranged from 1.15% to 2.46%. Again, the Freundlich equation fit the adsorption isotherms well, with the values of the exponent n around unity. The desorbed amount of lambda-cyhalothrin after five steps of desorption using deionized water was only 4.68% of the original amount, indicating that lambda-cyhalothrin adsorption to soil was virtually irreversible in water (Ali and Baugh 2003).

The phase distribution of lambda-cyhalothrin was investigated as a function of contact time in sediment (Bondarenko et al. 2006). For freshwater sediments at 9 d, the dissolved fraction measured by solid-phase microextraction ranged from 1.7% to 16.3% of the total pore-water concentration as determined by liquid–liquid extraction. The dissolved fraction decreased substantially with contact time to less than 5% at 30 d after sediment dosing. The dissolved fraction was lower in the marine sediment, ranging from 1.1% to 4.2%. Consequently, the apparent K_{OC} and dissolved organic carbon partition coefficient (K_{DOC}) values increased significantly over the contact time, especially in the freshwater sediment, suggesting that phase distribution was not at equilibrium after 9 d.

In general, the adsorption capacity of all soils with different organic content was much higher than that of pure mineral particles, suggesting that lambda-cyhalothrin had partitioned into the organic carbon phase as well as adsorbed onto the particle surfaces. However, in investigations of the adsorptive capacity of organic coatings on minerals, Zhou et al. (1995) discovered that adsorption was not constant and that it decreased with increasing polarity and decreasing aromaticity of organic sorbents. This result implied that the quantity, as well as the quality, of the organic matter on particle surfaces should be characterized to better predict contaminant transport in aquatic and soil environments.

4.2.2 Effect of pH on Lambda-Cyhalothrin Adsorption

The effect of pH on sorption was examined by conducting experiments over a range of four different pH values: 2, 4, 6, or 9 (Ali and Baugh 2003). The results showed that, over this range, pH did not have a significant impact, although soil-sorbed lambda-cyhalothrin at pH <4 was slightly higher than at pH 9. The results support the assumption that the adsorption of lambda-cyhalothrin is not significantly affected by changing the surface charge density of soil or organic matter.

4.2.3 Terrestrial Dissipation Studies

The dissipation of lambda-cyhalothrin in laboratory-controlled soils was mainly through biodegradation, as indicated by the rapid loss of lambda-cyhalothrin in nonsterile soil when compared to a sterile soil (Wang et al. 1997). However, the shape of the dissipation curve potentially indicated that processes primarily responsible for degradation differed over time. In the cotton soil, dissipation of lambda-cyhalothrin was fast in the first few days, which could indicate that degradation was initially through photolysis and hydrolysis chemical processes. The subsequent dissipation rate showed a steady slow decline, which could be evidence of microbial degradation. Dissipation curves for a Brazilian Oxisol-Typic Haplustox can be described by biexponential functions for lambda-cyhalothrin (Laabs et al. 2000). Soil in pans was treated with lambda-cyhalothrin at 15 g/ha using an indoor track sprayer, and the pans were dug into adjacent fallow and cropped areas of a field. Overall, the initial lambda-cyhalothrin residues (32 ppb or 11.8 g/ha) dissipated with a DT_{50} of 1.3 wk and a DT_{90} of 14.5 wk (disappearance time for first 50% and 90% of residue, respectively). Among treatments, residues in the soil pans shaded by the crop canopy declined faster (DT_{90} of 12.8 wk) than residues in the fallow area (DT_{90} of 16.2 wk). Compared with bare fallow, soil surface temperatures on warm, summer days were 8°–16°C cooler within the crop canopy. Also, the surface 0–2.5 cm of soil within the canopy took longer to dry out after a rainfall. It was hypothesized that the more ideal soil temperature and moisture conditions within the crop canopy increased the microbial degradation of lambda-cyhalothrin. One year after application, only 3.2% of the initial residues were recovered in the fallow area (Hill and Inaba 1991).

The influence of temperature and moisture on the degradation and persistence of lambda-cyhalothrin was investigated using laboratory incubation and lysimeters on a sandy loam soil (Typic Ustocurepts) in Pakistan (Tariq et al. 2006). Drainage from the lysimeters was sampled on days 49, 52, 59, 73, 100, 113, and 119 against the pesticide application on days 37, 63, 82, 108, and 137 after the sowing of cotton. The dissipation of lambda-cyhalothrin followed second-order kinetics. The results of incubation studies showed that increasing temperature and moisture content significantly reduced the $t_{1/2}$ values of pesticides in laboratory-controlled soils (Tariq et al. 2006; Wang et al. 1997).

Hydrosoil (sediment) appeared to act as a sink for lambda-cyhalothrin (Hadfield et al. 1993). Under the stringent test conditions of the mesocosm study, lambda-cyhalothrin residues in the hydrosoil (1.1% organic matter) reached 3.2 μg/kg following the seasonal exposure. Residues in the hydrosoil reached a maximum level of approximately 25 μg/kg in one sampling zone at one interval and thereafter declined to a level of 9 μg/kg within 4 mon.

4.2.4 Transport in Soil

In a study to evaluate the leaching potential of eight pesticides in a Brazilian Oxisol, lambda-cyhalothrin was applied onto a Typic Haplustox that contained ~50% clay and 26.3 g/kg organic carbon in the top 10-cm soil layer (Laabs et al. 2000). Mobility within the soil profile and subsequent leaching were studied for a period of 28 d after application. The bulk of lambda-cyhalothrin residues were recovered within the top 15 cm of the soil. In lysimeter percolates collected at 35 cm soil depth, less than 0.03% of the applied amount of lambda-cyhalothrin was recovered. The relative contamination potential of pesticides, according to the lysimeter study results, was ranked as follows: metolachlor > atrazine = simazine >> monocrotofos > endosulfan > chlorpyrifos > trifluralin > lambda-cyhalothrin. This same order of contamination potential was achieved by ranking the pesticides according to their effective sorption coefficient (K_e), which is the ratio of K_{oc} to field-dissipation half-life.

The potential for pesticide transport in preferential flow in an Oxisol was investigated in a study where lambda-cyhalothrin was applied on to a Typic Haplustox (Reichenberger et al. 2002). After application, a tracer solution containing 5 g/L of the dye Brilliant Blue FCF and 0.015 M KBr was applied at a rate of 40 mm/d in duplicate experiments over a period of 3 d. The solution was applied using either a tension infiltrometer (3.3 cm tension) or manual irrigation with a watering can. The soil monoliths were then opened and the soil layers at 0–5, 5–10, 10–20, 20–30, and 30–40 cm were quantitatively removed. Although the highest concentrations of lambda-cyhalothrin were found in the top 0–10 cm layer, detection of lambda-cyhalothrin in the deeper 10–30 cm soil layer was determined to result from preferential flow (Reichenberger et al. 2002; Tariq et al. 2006).

5 Ecotoxicity of Lambda-Cyhalothrin

Lambda-cyhalothrin is slightly to highly toxic to terrestrial and aquatic organisms. This review focuses on aquatic toxicity because aquatic organisms are most likely to be exposed to lambda-cyhalothrin residue levels found in water and associated sediment. It is worth noting that the reported oral LD_{50} (48 hr) for mallard duck was >3950 mg/kg, and that the oral LD_{50} (48 hr) was 0.038 μg/bee and the contact LD_{50} (48 h) was 0.909 μg/bee, indicating relatively low toxicity to ducks but high toxicity to bees (European-Commission 2001).

5.1 Fish and Shellfish

Lambda-cyhalothrin is highly toxic to a number of fish and shellfish. The reported LC_{50} (96 hr) is 210 ng/L for bluegill sunfish, 240 ng/L for rainbow trout, 360 ng/L for *Daphnia magna*, 4.9 ng/L for mysid shrimp, and 0.8 ng/L for sheepshead minnow. An EC_{50}, the concentration at which the effect occurs in 50% of the test population, for eastern oyster is 0.59 ng/L. A bioconcentration factor (BCF) of 2240 has been reported in fish (species unspecified), but concentration was confined to nonedible tissues and rapid depuration was observed (USDA 2007; USEPA 2007).

Because lambda-cyhalothrin is commonly applied to rice fields to control insects, potential water and sediment contamination may lead to toxicity in aquatic organisms such as mosquitofish, shrimps, crabs, and clams. Replicated enclosures in a rice field were sprayed with the lambda-cyhalothrin product Warrior at 5.8 g a.i./ha. Mosquitofish were added either before the spray or 7 d later. Of those added before the spray, none survived. Most fish added 7 d later survived (Lawler et al. 2003). Lambda-cyhalothrin showed high toxicity to shrimp (*Macrobrachium nippoensis* de Haan) and zebrafish (*Brachydanio rerio* H.B). The 96-hr LC_{50} was 20–70 ng/L for shrimp and 0.98–7.55 μg/L for zebrafish. In drainage water ponds with lambda-cyhalothrin concentrations ranging from 0.45 to 0.90 μg/L, the 96-hr mortality was 100% for shrimp, but the drainage water showed no toxicity to shrimp on the fourth day after application of lambda-cyhalothrin (Gu et al. 2007). The 96-hr LC_{50} values for lambda-cyhalothrin and gamma-cyhalothrin were similar for zebrafish (1.93 μg/L for gamma and 1.94 μg/L for lambda). However, lambda-cyhalothrin was more toxic (LC_{50}, 0.04 μg/L) than gamma-cyhalothrin (LC_{50}, 0.28 μg/L) to shrimp, possibly implying that the toxicity to shrimp is likely stereochemistry dependent (Wang et al. 2007).

5.2 *Macrophytes*

The structure of an ecosystem determines the final effect of pesticide exposure to macrophytes (Wendt-Rasch et al. 2004). Using a pesticide mixture containing lambda-cyhalothrin applied to 10 mesotrophic aquatic ecosystems dominated by submerged macrophytes (*Elodea*) and 10 simulated eutrophic ecosystems with a high *Lemna* surface coverage, significant increases in the biomass and alterations of species composition of the periphytic algae were observed in the *Elodea*-dominated microcosms, but no effect on *Myriophyllum spicatum* growth was observed. The opposite was found in the *Lemna*-dominated microcosms, in which decreased growth of *M. spicatum* was observed but no alterations were observed in the periphytic community. Furthermore, application of 0.17 and 1.7 g a.i. ha^{-1} lambda-cyhalothrin to pond mesocosms failed to produce adverse effects on macrophytes (Farmer et al. 1995).

5.3 *Invertebrates*

Lambda-cyhalothrin effects on mesotrophic (macrophyte-dominated) and eutrophic (phytoplankton-dominated) ditch microcosms were studied; applications were made three times at 1-wk intervals and at concentrations of 10, 25, 50, 100, and

250 ng/L (Roessink et al. 2005). Initial, direct effects were primarily on arthropod taxa. The most sensitive species was the phantom midge (*Chaoborus obscuripes*). At treatment levels of 25 ng/L and higher, apparent population and community responses occurred. At treatments of 100 and 250 ng/L, the rate of recovery of the macroinvertebrate community was lower in the macrophyte-dominated systems, primarily because of a prolonged decline of the amphipod *Gammarus pulex*. This species occurred at high densities only in the macrophyte-dominated enclosures. Indirect effects (e.g., increase of rotifers and microcrustaceans) were more pronounced in the plankton-dominated test systems, particularly at treatment levels of 25 ng/L and higher.

Simulated aquatic ditch mesocosms were used to understand the toxic effects of potential spray drift from a typical crop application of pesticides including lambda-cyhalothrin. Spray application to the water surface was at 0.2%, 1%, and 5% of the recommended label rates. To interpret the observed effects, treatment concentrations were expressed in toxic units (TU = actual concentration/EC_{50}) (Arts et al. 2006), which describe the relative toxicity of the compounds with common toxicity test organisms (*Daphnia* and algae). After treatment, lambda-cyhalothrin disappeared from the water phase within 2 d. At the 5% treatment level, exposure concentrations exceeded 0.1 $TU_{Daphnia}$, and this resulted in long-term effects on zooplankton and macroinvertebrates, some of which did not fully recover by the end of the study (Arts et al. 2006). Previous studies with pesticides in experimental ecosystems have demonstrated that effects on primary producers are likely to occur at TU_{Algae} >0.1 and effects on invertebrates are likely to occur at TU >0.01–0.1 (Van Wijngaarden et al. 2004, 2005).

5.4 Sediment Toxicity

Recent studies have shown that pyrethroids including lambda-cyhalothrin are commonly found in aquatic sediments in the heavily agricultural Central Valley of California, and therefore the toxicity of sediment-associated pyrethroid residues to aquatic organisms has been actively investigated (Amweg et al. 2005; Weston et al. 2004). Seventy sediment samples were collected from 42 sites over a 10-county area in the agriculture-dominated Central Valley of California, with most sites located in irrigation canals and small creeks dominated by agricultural effluent. Significant mortality was observed with the amphipod *Hyalella azteca* at 42% of the locations and for the midge *Chironomus tentans* at 40% of the sites. Using a toxicity unit analysis, measured pyrethroid concentrations were sufficiently high to have contributed to the toxicity in 40% of samples toxic to *C. tentans* and nearly 70% of samples toxic to *H. azteca* (Weston et al. 2004).

In a follow-up aquatic toxicity study with six pyrethroids (Amweg et al. 2005), the average 10-d median lethal concentration LC_{50} of sediment-associated residues of lambda-cyhalothrin was 0.45 μg/g OC (organic carbon adjusted), corresponding to an estimated pore-water concentration of 1.4 ng/L. Lambda-cyhalothrin would be acutely toxic to *H. azteca* at the concentration of 5.6 ng/g dry wt of sediment. Growth was typically inhibited at concentrations below the LC_{50}; animal biomass on average was 38% below controls when exposed to lambda-cyhalothrin concentrations roughly one-third to one-half the LC_{50}; i.e., the growth lowest-observable-effect concentration (LOEC) was 0.19 µg/g OC. Survival data indicate that exposure occurs primarily via the interstitial water rather than the particulate phase.

If only the dissolved concentration is bioavailable (Gan et al. 2006; Maund et al. 1998), these observations suggest that contact time after sediment dosing may greatly affect the bioavailability and, hence, the toxicity of pyrethroids. Therefore, a long contact time (30 d) is recommended for sediment toxicity testing of this class of compounds. The dependence of bioavailability on contact time also implies that test conditions must be standardized to allow comparison between laboratory-dosed samples and field samples (Bondarenko et al. 2006).

5.5 *Effects on Soil Fauna*

Species sensitivity distributions (SSD) and 5% hazardous concentrations (HC5) are distribution-based approaches for assessing environmental risks of pollutants, e.g., lambda-cyhalothrin risks to soil invertebrate communities. From a systematic review of literature, a total of 1950 laboratory toxicity test results were obtained, representing 250 pesticides including lambda-cyhalothrin and 67 invertebrate taxa. The majority (96%) of pesticides have toxicity data on fewer than 5 species. Based on a minimum of 5 species, the best available endpoint data (acute mortality median lethal concentration) enabled SSD and HC5 to be calculated for 11 pesticides including lambda-cyhalothrin. Arthropods and oligochaetes exhibit pronounced differences in their sensitivity to most of these pesticides. The standard test earthworm species, *Eisenia fetida sensu lato*, is least sensitive to insecticides based on acute mortality, whereas the standard *Collembola* test species, *Folsomia candida*, is among the most sensitive species for a broad range of toxic modes of action (biocide, fungicide, herbicide, and insecticide) (Frampton et al. 2006). To assess the effects of lambda-cyhalothrin on soil invertebrates under tropical conditions, ecotoxicological semifield studies were conducted using intact soil-core terrestrial model ecosystems (TMEs) (Forster et al. 2006). Earthworms, isopods, and diplopods were added to intact soil cores and the mortality of soil invertebrates was determined. The results indicated that lambda-cyhalothrin was toxic to isopods and millipedes, whereas no effect on arthropods was detected in the field.

To evaluate possible microbial community changes in a sandy loam soil in response to the addition of lambda-cyhalothrin, the following properties were determined: active soil microbial biomass, concentrations of ammonium and nitrate ions, numbers of total cultivable bacteria, fungi, nitrogen-fixing bacteria, and nitrifying and denitrifying bacteria (Cycon et al. 2006). Substrate-induced respiration (SIR) increased with time in controls ranging from 13.7 to 23.7 mg O_2/

kg dry soil/hr and in pesticide-treated soil ranging from 12–13 to 23–25 O_2/kg dry soil/hr on days 1 and 28, respectively. The concentrations of nitrate and ammonium ions, numbers of total cultivable bacteria, denitrifying bacteria, nitrogen-fixing bacteria, and fungi were either unaffected or even stimulated by the pesticide treatments.

5.5 *Bioaccumulation*

The interaction of lambda-cyhalothrin with larvae of the aquatic insect *Chironomus riparius* was studied in laboratory sediment–water systems (Hamer et al. 1999). *C. riparius,* a nonbiting midge, is widely used for investigating toxicity and bioaccumulation of contaminants in sediment. Ten different sediments were used. ^{14}C-Labeled lambda-cyhalothrin was applied to sediment slurries and *C. riparius* was exposed in the test system for 48 hr. In all the sediment–water test systems, >99% of the ^{14}C-labeled lambda-cyhalothrin was adsorbed onto the sediment. BCFs based on the aqueous-phase concentrations showed little difference among systems, ranging from 1,300 to 3,400 with a mean of 2,300 and with a coefficient of variation of 25%. These values were very similar to a BCF of 2,000 that was determined in water alone after 48 hr. These BCF values are similar to chlorpyrifos, which is 1,374, but higher than diazinon, which is 500 (PAN 2007). BCFs based on measured concentrations of extractable ^{14}C-lambda-cyhalothrin in the sediment phase were always < 1, ranging from 0.11 to 0.84 with a mean of 0.39 and coefficient of variation of 61%. Sediment BCFs were inversely proportional to the measured sediment K_ds, which ranged from 3,290 to 22,100; that is, the higher the proportion of the chemical that was adsorbed, the lower the sediment BCF. The results of the study supported equilibrium partitioning theory. In sediment–water systems, the lambda-cyhalothrin that was bioavailable is equivalent to the amount that is measured in the water phase.

6 Mitigating Runoff of Residues Through Plant Interaction

Because lambda-cyhalothrin residues in water and sediment may exert adverse impacts on aquatic life, management practices have been investigated and developed to mitigate such impact. Plants have long been recognized to have the ability to sequester a variety of contaminants (e.g., organic or inorganic chemicals) in water, sediment, and soil (Bouldin et al. 2006; Burken and Schnoor 1997; Henderson et al. 2007; Mertens et al. 2006; Montes-Bayon et al. 2002; Siciliano et al. 1998; Wild et al. 2005). In recent years, the efficacy of vegetated ditches to reduce pesticide residues in agricultural runoff has been extensively studied. The results indicate that vegetated ditches are useful for the reduction of lambda-cyhalothrin in stormwater or agricultural runoff (Bennett et al. 2005; Bouldin et al. 2005; Leistra et al. 2004; Moore et al. 2001; Roessink et al. 2005). Pesticide uptake, adsorption, and accelerated degradation by plants in a vegetated ditch are the principal mechanisms of mitigation by plants, as described next.

6.1 Assimilation by Plants

Macrophytes (e.g., *Juncus effuses* and *Ludwigia peploides*) can take up a significant portion of lambda-cyhalothrin residues in water. Eight days after application, 98.2% of lambda-cyhalothrin was found in the roots of *L. peploides*. Translocation of lambda-cyhalothrin in *J. effusus* resulted in 25.4% of pesticide uptake partitioning into upper plant biomass. These macrophytes showed species- and pesticide-specific uptake rates, and therefore the selection of high-uptake plants enhances mitigation capabilities in edge-of-field conveyance structures (Bouldin et al. 2006).

6.2 Adsorption and Degradation by Plants

Plants growing in an agricultural drainage ditch not only slow water flow and absorb chemicals dissolved in water, but they will also act as surfaces for pesticide adsorption and catalysts for their degradation. Many studies have been conducted to investigate the effectiveness of vegetated drainage ditches for reducing pesticide residue export from agricultural lands to creeks and rivers (Bennett et al. 2005; Bouldin et al. 2005; Moore et al. 2001). By incorporating vegetated drainage ditches into a watershed management program, agriculture can continue to decrease potential non-point source threats to downstream aquatic receiving systems. Overall results of previous studies illustrate that aquatic macrophytes play an important role in the retention and distribution of pyrethroids, including lambda-cyhalothrin, in vegetated agricultural drainage ditches.

The reduction effectiveness of pesticides through a drainage ditch is usually described by the percentage of reduction of the pesticide:

$$\text{Reduction\%} = (\text{conc. at inlet} - \text{conc. at outlet}) \times 100/\text{conc. at inlet}$$

The reduction% is affected by several factors including the length and slope of ditch, the type and density of vegetation, and the type of soil (Arts et al. 2006; Bennett et al. 2005; Bouldin et al. 2005; Leistra et al. 2004; Milam et al. 2004; Moore et al. 2001; Roessink et al. 2005).

Laboratory studies indicated that adsorption to macrophytes was extensive and essentially irreversible, and that degradation of lambda-cyhalothrin occurred rapidly by cleavage of the ester bond. In the indoor microcosm, which contained water, sediment, and macrophytes, degradation was also rapid, with DT_{50} and DT_{90} values of less than 3 and 19 hr, respectively, for dissipation from the water column and less than 3 and 56 hr, respectively, for dissipation from the whole system (Hand et al. 2001). The adsorption of lambda-cyhalothrin by *Ludwigia peploides* (water primrose) and *Juncus effusus* (soft rush) was significant and reached as high as 86.50 µg/kg (Bouldin et al. 2005).

In a vegetated drainage ditch study, following initiation of simulated runoff, mean percentages of lambda-cyhalothrin concentrations in water and sediment in the ditch were 12% and 1%, respectively. Lambda-cyhalothrin mean percentage concentrations in plants [*Polygonum* (water smartweed), *Leersia* (cutgrass), and *Sporobolus* (smutgrass)] were 87%. The concentrations in water decreased to levels safe for nontarget species 50 m downstream from the point of input (Moore et al. 2001).

A controlled-release runoff simulation was conducted on a 650-m vegetated drainage ditch in the Mississippi Delta. Lambda-cyhalothrin was released into the ditch in a water–sediment slurry (Bennett et al. 2005). Samples of water, sediment, and plants were collected and analyzed for lambda-cyhalothrin concentrations. Three hours following runoff initiation, inlet lambda-cyhalothrin water concentration was 374 µg/L in the inlet and 7.24 µg/L at 200 m downstream. No lambda-cyhalothrin residues were detected at the 400-m sampling site. A similar trend was observed throughout the first 7 d of the study where water concentrations were elevated at the front end of the ditch (~25 m) and greatly reduced downstream at the 400-m sampling site. Regression formulas predicted that lambda-cyhalothrin concentrations in ditch water were reduced to 10% of the initial value within 280 m downstream from point of input. Mass balance calculations determined that ditch plants constituted the major sink and/or sorption site responsible for the rapid aqueous pyrethroid dissipation.

In a ditch study conducted in the Netherlands, concentrations of lambda-cyhalothrin decreased rapidly in the water column: 24%–40% of the dose remained in the water 1 d after application, and it had decreased to 1.8%–6.5% after 3 d (Leistra et al. 2004). At the highest plant density, lambda-cyhalothrin residues in the plant compartment reached a maximum of 50% of the dose after 1 d; at intermediate and low plant densities, this maximum was only 3%–11% of the dose (after 1–2 d). The percentage of the insecticide in the ditch sediment was 12% or less of the dose and tended to be lower at higher plant densities. Alkaline hydrolysis in the water near the surface of macrophytes and phytoplankton is considered to be the main dissipation process for lambda-cyhalothrin.

7 Summary

Lambda-cyhalothrin is a pyrethroid insecticide used for controlling pest insects in agriculture, public health, and in construction and households. Lambda-cyhalothrin is characterized by low vapor pressure and a low Henry's law constant but by a high octanol–water partition coefficient (K_{ow}) and high water–solid-organic carbon partition coefficient (K_{oc}) values. Lambda-cyhalothrin is quite stable in water at pH < 8, whereas it hydrolyzes to form HCN and aldehyde under alkaline conditions. Although lambda-cyhalothrin is relatively photostable under natural irradiation, with a half-life > 3 wk, its photolysis process is fast under UV irradiation, with a half-life < 10 min. The fate of lambda-cyhalothrin in aquatic ecosystems depends on the nature of system components such as suspended solids (mineral and organic particulates) and aquatic organisms (algae, macrophytes, or aquatic animals). Lambda-cyhalothrin residues dissolved in water decrease rapidly if suspended solids and/or aquatic organisms are present because lambda-cyhalothrin molecules are strongly adsorbed by particulates and plants. Adsorbed lambda-cyhalothrin molecules show decreased degradation rates because they are less accessible to breakdown than free molecules in the water column. On the other hand, lambda-cyhalothrin adsorbed to suspended solids or bottom sediments may provide a mechanism to mitigate its acute toxicity to aquatic organisms by reducing their short-term bioavailability in the water column. The widespread use of lambda-cyhalothrin has resulted in residues in sediment, which have been found to be toxic to aquatic organisms including fish and amphipods. Mitigation measures have been used to reduce the adverse impact of lambda-cyhalothrin contributed from agricultural or urban runoff. Mitigation may be achieved by reducing the quantity of runoff and suspended solid content in runoff through wetlands, detention ponds, or vegetated ditches.

References

Ali MA, Baugh PJ (2003) Sorption-desorption studies of six pyrethroids and mirex on soils using GC/MS-NICI. Int J Environ Anal Chem 83:923–933.

Amweg EL, Weston DP (2005) Use and toxicity of pyrethroid pesticides in the Central Valley, California, USA. Environ Toxicol Chem 24:1300–1301.

Amweg EL, Weston DP, Ureda NM (2005) Use and toxicity of pyrethroid pesticides in the Central Valley, California, USA. Environ Toxicol Chem 24:966–972.

Amweg EL, Weston DP, You J, Lydy MJ (2006) Pyrethroid insecticides and sediment toxicity in urban creeks from California and Tennessee. Environ Sci Technol 40:1700–1706.

Arts GHP, Guijse-Bogdan LL, Belgers JDM, Van Rhenen-Kersten CH, Van Wijngaarden RPA, Roessink I, Maund SJ, Van den Brink PJ, Brock TCM (2006) Ecological impact in ditch mesocosms of simulated spray drift from a crop protection program for potatoes. Integr Environ Assess Manag 2:105–125.

Bennett ER, Moore MT, Cooper CM, Smith S, Shields FD, Drouillard KG, Schulz R (2005) Vegetated agricultural drainage ditches for the mitigation of pyrethroid-associated runoff. Environ Toxicol Chem 24:2121–2127.

Bondarenko S, Putt A, Kavanaugh S, Poletika N, Gan JY (2006) Time dependence of phase distribution of pyrethroid insecticides in sediment. Environ Toxicol Chem 25:3148–3154.

Bouldin JL, Farris JL, Moore MT Jr, Stephens WW, Cooper CM (2005) Evaluated fate and effects of atrazine and lambda-cyhalothrin in vegetated and unvegetated microcosms. Environ Toxicol 20:487–498.

Bouldin JL, Farris JL, Moore MT, Smith JS, Cooper CM (2006) Hydroponic uptake of atrazine and lambda-cyhalothrin in *Juncus effusus* and *Ludwigia peploides*. Chemosphere 65:1049–1057.

Bradbury SP, Coats JR (1989) Toxicokinetics and toxicodynamics of pyrethroid insecticides in fish. Environ Toxicol Chem 8:373–380.

Burken JG, Schnoor JL (1997) Uptake and metabolism of atrazine by poplar trees. Environ Sci Technol 31:1399–1406.

Burr SA, Ray DE (2004) Structure-activity and interaction effects of 14 different pyrethroids on voltage-gated chloride ion channels. Toxicol Sci 77:341–346.

Cavas T, Ergene-Gozukara S (2003) Evaluation of the genotoxic potential of lambda-cyhalothrin using nuclear and nucleolar biomarkers on fish cells. Mutat Res/Genet Toxicol Environ Mutagen 534:93–99.

CDPR (2006) California Department of Pesticide Regulation Pesticide Use Database.

CDPR (2007) Pesticide chemical database.

Cleugh ES, Milner DJ (1994) Isomerization process. http://wwwpatentstormus/patents/5334744-descriptionhtml.

Cycon M, Piotrowska-Seget Z, Kaczynska A, Kozdroj J (2006) Microbiological characteristics of a sandy loam soil exposed to tebuconazole and lambda-cyhalothrin under laboratory conditions. Ecotoxicology 15:639–646.

European-Commission (2001) Review report for the active substance lambda-cyhalothrin. 7572/VI/97-final.25January2001.http://ec.europa.eu/food/plant/protection/evaluation/existactive/list1-24_en.pdf.

Farmer D, Hill IR, Maund SJ (1995) A comparison of the fate and effects of two pyrethroid insecticides (lambda-cyhalothrin and cypermethrin) in pond mesocosms. Ecotoxicology 4:219–244.

Fernandez-Alvarez M, Sanchez-Prado L, Lores M, Llompart M, Garcia-Jares C, Cela R (2007) Alternative sample preparation method for photochemical studies based on solid phase micro-extraction: synthetic pyrethroid photochemistry. J Chromatogr A Adv Sample Prep 1152:156–167.

Forster B, Garcia M, Francimari O, Rombke J (2006) Effects of carbendazim and lambda-cyhalothrin on soil invertebrates and leaf litter decomposition in semi-field and field tests under tropical conditions (Amazonia, Brazil). Eur J Soil Biol 42:S171–S179.

Frampton GK, Jansch S, Scott-Fordsmand JJ, Rombke J, Van den Brink PJ (2006) Effects of pesticides on soil invertebrates in laboratory studies: a review and analysis using species sensitivity distributions. Environ Toxicol Chem 25:2480–2489.

Gan J, Yang W, Hunter W, Bondarenko S, Spurlock F (2006) Bioavailability of pyrethroids in surface aquatic systems. http://wwwcdprcagov/docs/sw/presentations/JGan_pyrethroids 101105pdf.

Gu BG, Wang HM, Chen WL, Cai DJ, Shan ZJ (2007) Risk assessment of lambda-cyhalothrin on aquatic organisms in paddy field in China. Regul Toxicol Pharmacol 48:69–74.

Gupta S, Handa SK, Sharma KK (1998) A new spray reagent for the detection of synthetic pyrethroids containing a nitrile group on thin-layer plates. Talanta 45:1111–1114.

Hadfield ST, Sadler JK, Bolygo E, Hill S, Hill IR (1993) Pyrethroid residues in sediment and water samples from mesocosm and farm pond studies of simulated accidental aquatic exposure. Pestic Sci 38:283–294.

Hamer MJ, Goggin UM, Muller K, Maund SJ (1999) Bioavailability of lambda-cyhalothrin to *Chironomus riparius* in sediment-water and water-only systems. Aquat Ecosys Health Manag 2:403–412.

Hand LH, Kuet SF, Lane MCG, Maund SJ, Warinton JS, Hill IR (2001) Influences of aquatic plants on the fate of the pyrethroid insecticide lambda-cyhalothrin in aquatic environments. Environ Toxicol Chem 20:1740–1745.

Heckmann LH, Friberg N (2005) Macroinvertebrate community response to pulse exposure with the insecticide lambda-cyhalothrin using in-stream mesocosms. Environ Toxicol Chem 24:582–590.

Henderson KL, Belden JB, Coats JR (2007) Mass balance of metolachlor in a grassed phytoremediation system. Environ Sci Technol 41:4084–4089.

Hill BD, Inaba DJ (1991) Dissipation of lambda-cyhalothrin on fallow vs. cropped soil. J Agric Food Chem 39:2282–2284.

Laabs V, Amelung W, Pinto A, Altstaedt A, Zech W (2000) Leaching and degradation of corn and soybean pesticides in an Oxisol of the Brazilian Cerrados. Chemosphere 41:1441–1449.

Laskowski DA (2002) Physical and chemical properties of pyrethroids. Rev Environ Contam Toxicol 174:49–170.

Lawler SP, Dritz DA, Godfrey LD (2003) Effects of the agricultural insecticide lambda-cyhalothrin (Warrior (™)) on mosquitofish (*Gambusia affinis*). J Am Mosq Control Assoc 19:430–432.

Lawler SP, Dritz DA, Christiansen JA, Cornel AJ (2007) Effects of lambda-cyhalothrin on mosquito larvae and predatory aquatic insects. Pest Manag Sci 63:234–240.

Leistra M, Zweers AJ, Warinton JS, Crum SJ, Hand LH, Beltman WH, Maund SJ (2004) Fate of the insecticide lambda-cyhalothrin in ditch enclosures differing in vegetation density. Pest Manag Sci 60:75–84.

Maund SJ, Hamer MJ, Warinton JS, Kedwards TJ (1998) Aquatic ecotoxicology of the pyrethroid insecticide lambda-cyhalothrin: considerations for higher-tier aquatic risk assessment. Pestic Sci 54:408–417.

Mertens J, Vervaeke P, Meers E, Tack FMG (2006) Seasonal changes of metals in willow (*Salix* sp.) stands for phytoremediation on dredged sediment. Environ Sci Technol 40:1962–1968.

Milam CD, Bouldin JL, Farris JL, Schulz R, Moore MT, Bennett ER, Cooper CM, Smith S (2004) Evaluating acute toxicity of methyl parathion application in constructed wetland mesocosms. Environ Toxicol 19:471–479.

Montes-Bayon M, Yanes EG, Ponce de Leon C, Jayasimhulu K, Stalcup A, Shann J, Caruso JA (2002) Initial studies of selenium speciation in *Brassica juncea* by LC with ICPMS and ES-MS detection: an approach for phytoremediation studies. Anal Chem 74:107–113.

Moore MT, Bennett ER, Cooper CM, Smith S, Shields FD, Milam CD, Farris JL (2001) Transport and fate of atrazine and lambda-cyhalothrin in an agricultural drainage ditch in the Mississippi Delta, USA. Agric Ecosyst Environ 87:309–314.

Oros DR, Werner I (2005) Pyrethroid Insecticides: An Analysis of Use Patterns, Distributions, Potential Toxicity and Fate in the Sacramento-San Joaquin Delta and Central Valley. White Paper for the Interagency Ecological Program. SFEI Contribution 415. San Francisco Estuary Institute, Oakland, CA.

Oudo H, Hansen HC (2002) Sorption of lambda-cyhalothrin, cypermethrin, deltamethrin and fenvalerate to quartz, corundum, kaolinite and montmorillonite. Chemosphere 49:1285–1294.

PAN (2007) PAN Pesticides Database, Chemicals. http://www.pesticideinfo.org/List_Chemicals.jsp. Accessed July 20, 2007.

Reichenberger S, Amelung W, Laabs V, Pinto A, Totsche KU, Zech W (2002) Pesticide displacement along preferential flow pathways in a Brazilian Oxisol. Geoderma 110:63–86.

Robson MJ and Crosby J (1984) Insecticidal product and preparation thereof. European Patent Office. Patent Number EU 106469. UK.

Roessink I, Arts GHP, Belgers JDM, Bransen F, Maund SJ, Brock TCM (2005) Effects of lambda-cyhalothrin in two ditch microcosm systems of different trophic status. Environ Toxicol Chem 24:1684–1696.

Ruzo LO, Krishnamurthy VV, Casida JE, Gohre K (1987) Pyrethroid photochemistry: influence of the chloro(trifluoromethyl)vinyl substituent in cyhalothrin. J Agric Food Chem 35:879–883.

Schwarzenbach RP, Gschwend PM, Imboden DM (1993) Environmental Organic Chemistry. Wiley, New York.

Shafer TJ, Meyer DA (2004) Effects of pyrethroids on voltage-sensitive calcium channels: a critical evaluation of strengths, weaknesses, data needs, and relationship to assessment of cumulative neurotoxicity. Toxicol Appl Pharmacol 196:303–318.

Siciliano SD, Goldie H, Germida JJ (1998) Enzymatic activity in root exudates of Dahurian wild rye (*Elymus dauricus*) that degrades 2-chlorobenzoic acid. J Agric Food Chem 46:5–7.

Spurlock F (2006) Synthetic pyrethroids and California surface water: use patterns, properties, and unique aspects. http://wwwcdprcagov/docs/sw/swposters/spurlock_acs06pdf.

Starner K (2007) Data queried from the Department of Pesticide Regulation Surface Water Monitoring Database.

Syngenta (2007) KARATE. http://wwwsyngentacom/en/products_services/karate_pageaspx.
Tariq MY, Afzal S, Hussain I (2006) Degradation and persistence of cotton pesticides in sandy loam soils from Punjab, Pakistan. Environ Res 100:184–196.
Tomlin CDS (ed) (2000) The Pesticide Manual, 12th Ed. British Crop Protection Council, Farnham, UK.
USDA (2007) USDA-ARS Pesticide Properties Database: http://www.ars.usda.gov/Services/docs.htm?docid=14199. Accessed July 20, 2007.
USEPA (2007) ECOTOX database. http://cfpubepagov/ecotox/quick_queryhtm.
Van Wijngaarden RPA, Cuppen JGM, Arts GHP, Crum SJH, van den Hoorn MW, Van den Brink PJ, Brock TCM (2004) Aquatic risk assessment of a realistic exposure to pesticides used in bulb crops: a microcosm study. Environ Toxicol Chem 23:1479–1498.
Van Wijngaarden RPA, Brock TCM, Van den Brink PJ (2005) Threshold levels for effects of insecticides in freshwater ecosystems: a review. Ecotoxicology 14:355–380.
Wang S, Kimber SWL, Kennedy IR (1997) The dissipation of lambda-cyhalothrin from cotton production systems. J Environ Sci Health B B32:335–352.
Wang W, Cai DJ, Shan ZJ, Chen WL, Poletika N, Gao XW (2007) Comparison of the acute toxicity for gamma-cyhalothrin and lambda-cyhalothrin to zebra fish and shrimp. Regul Toxicol Pharmacol 47:184–188.
Wendt-Rasch L, Van den Brink PJ, Crum SJH, Woin P (2004) The effects of a pesticide mixture on aquatic ecosystems differing in trophic status: responses of the macrophyte *Myriophyllum spicatum* and the periphytic algal community. Ecotoxicol Environ Saf 57:383–398.
Weston DP, You JC, Lydy MJ (2004) Distribution and toxicity of sediment-associated pesticides in agriculture-dominated water bodies of California's Central Valley. Environ Sci Technol 38:2752–2759.
Weston DP, Holmes RW, You J, Lydy MJ (2005) Aquatic toxicity due to residential use of pyrethroid insecticides. Environ Sci Technol 39:9778–9784.
Wild E, Dent J, Thomas GO, Jones KC (2005) Direct observation of organic contaminant uptake, storage, and metabolism within plant roots. Environ Sci Technol 39:3695–3702.
Zhou JL, Rowland S, Mantoura RFC (1995) Partition of synthetic pyrethroid insecticides between dissolved and particulate phases. Water Res 29:1023–1031.

Lead Contamination in Uruguay: The "La Teja" Neighborhood Case

Nelly Mañay, Adriana Z. Cousillas, Cristina Alvarez, and Teresa Heller

1 Introduction

Our last review, "Lead Contamination in Uruguay" (Mañay et al. 1999), provided essential background information for the lead contamination incident in the La Teja neighborhood of Montevideo, Uruguay. Our reports, including those cited in this review, were officially acted upon by Sanitary and Environmental authorities beginning in late 2000 (Mañay et al. 2003). Before the release of our review, information on lead contamination in Uruguay was dispersed and incomplete. Although health risks to residents associated with lead contamination and exposure in polluted areas

N. Mañay(✉)
Department of Toxicology and Environmental Hygiene, Faculty of Chemistry, University of the Republic of Uruguay, Gral. Flores 2124, 11800 Montevideo, Uruguay(nmanay@fq.edu.uy)

A.Z. Cousillas
Department of Toxicology and Environmental Hygiene, Faculty of Chemistry, University of the Republic of Uruguay, Gral. Flores 2124, 11800 Montevideo, Uruguay

C. Alvarez
Department of Toxicology and Environmental Hygiene, Faculty of Chemistry, University of the Republic of Uruguay, Gral. Flores 2124, 11800 Montevideo, Uruguay

T. Heller
Department of Toxicology and Environmental Hygiene, Faculty of Chemistry, University of the Republic of Uruguay, Gral. Flores 2124, 11800 Montevideo, Uruguay

D.M. Whitacre (ed.), *Reviews of Environmental Contamination and Toxicology.*

often went unrecognized, contamination events were previously known. An example is the Malvín Norte neighborhood case described by Cousillas et al. (1998).

The Mañay et al. (1999) review was considered both interesting and useful to the scientific community, to the press, and to the general public. This review, together with associated scientific research results published by Mañay (2001a,b), by Schutz et al. (1997), or presented at scientific meetings, substantively contributed to recognition of lead poisoning as a rising health problem in Uruguay. Concomitantly, our research team also contributed its knowledge of lead blood analysis and biomonitoring on different Uruguayan populations to others working on the problem.

It was not until 2001 that official attention was increasingly being paid to environmental lead exposure, although our research group had published several studies during the 1990s, including those reviewed in Mañay et al. (1999). The understanding of environmental risks associated with lead contamination in Uruguay improved as data on lead concentrations in blood and soil samples accumulated. The growing scientific evidence, along with press reports and court cases, increased awareness and concern among Uruguayans for lead-induced health risks (Amorin 2001; Matos 2001). As a consequence, new research studies on Uruguayan populations became available, including some that had gone unpublished for several years. Some interdisciplinary study results were officially communicated to the public as they became available, particularly those designed to prevent environmental health risks to children.

To illustrate the growing momentum for change, in late 2003, the use of tetraethyl lead in gasoline was phased out in Uruguay. Although not as yet officially evaluated, this change was expected to reduce lead levels in air and consequently reduce human blood lead levels as well.

2 Lead Exposure in Uruguay

The adverse effects of lead are well known, environmentally exposed children being the most affected population. Lead exposure is known to threaten human health, and, although lead has no known biological role, it is frequently found as a residue in human tissues. There is a significant negative correlation between the mental development of children and environmental lead exposure (ACHS 2000; ATSDR 2003, 2005; IPCS 1995).

2.1 Sources of Lead

Uruguay is a small country comprising 3 million inhabitants, half of whom live in the coastal city of Montevideo, its capital. Most lead processing industries in Uruguay, such as metallurgies, foundries, manufacturers of batteries, and battery recyclers, are established in or around Montevideo, thus contributing to the contamination of the peripheral slums.

Until December 2003, gasoline used in Uruguay contained tetraethyl lead (TEL) as an antiknock additive in concentrations ranging from 150 to 300 mg/L (Barañano et al. 2005). Use of TEL in Uruguay was replaced in 2004 by methyl *tert*-butyl ether.

Montevideo still has several lead-emitting industries that pose some continuing exposure risk to people, and most of these are located in residential areas. Another source of population exposure to lead is through the use of lead pipes to conduct drinking water to and within old buildings. Individuals who live near manufacturing areas that do not handle lead materials or waste safely are also potentially at risk (Dol et al. 2004; IMM 2003; Mañay et al. 2003). Other potential sources of exposure, although not yet studied, include lead-based paint in old buildings and lead residues released from smoking tobacco (UNEP 2007; UNEP/OECD 1999).

2.2 Environmental Lead Exposure: The La Teja Case (Montevideo)

In 2001, lead exposure became a matter of public concern in Uruguay when several cases of children with blood lead levels (BLL) higher than 20 μg/dL were discovered. Burger and Pose (2001) described the first such case, a child with BLL of 31.2 μg/dL; after this, many more high BLL cases were discovered. The affected children were residents of a low-income neighborhood in Montevideo called La Teja (Mañay et al. 2003). It was reported that after lead handling factories were abandoned, many families settled in the area (La Teja) without either knowing it was polluted with lead or understanding the high risks to themselves or their children from living on this site.

In 2001, official lead surveillance analyses began on residents' blood and on environmental samples collected from those abandoned industrial areas (Ponzo 2002). The highest lead levels were found in soil samples from some slum settlements; residue levels exceeded 3000 ppm and resulted from lead scrap landfills (Dol et al. 2004; IMM 2003).

Uruguay had encountered many previous lead contamination episodes (Ascione 2001; Cousillas et al. 1998). However, in early 2001, the "Lead in La Teja" case became unique in Uruguay because it developed into an environmental, sanitary, and social emergency. As a result, an official multidisciplinary expert group with State University and community delegates was created. The serious nature of this pollution event resulted in an initiative by the Ministry of Health to hold periodic meetings and to undertake needed remediation.

Ascione (2001) pointed out that soil and dust are important sources of lead exposure for children. This article stressed the importance of pediatricians evaluating their patients for possible lead intoxication and following up by taking the proper clinical steps if symptoms were compatible with lead poisoning. Lead poisoning symptoms include behavioral disturbances, intellectual deficit, hyperactivity, falling behind in school, renal pathology, etc.

In Mañay et al. (2003), a preliminary approach to the environmental lead pollution problem in the La Teja neighborhood of Montevideo was presented through the appearance of high blood lead levels in children living there. These early BLL results were first discussed in a Latin-American lead workshop held in Lima, Perú (Mañay 2001a), and those data were obtained by analyzing venous BLL in children and pregnant women using atomic absorption spectrometry (AAS). Lead analyses on environmental samples were performed at the same time so as to identify the possible lead sources (e.g., drinking water, soil, and air). These data were collected within the framework of the Interinstitutional Lead Committee assembled in 2001.

The aforementioned paper (Mañay et al. 2003) also reported the onset of periodic health care visits to children who had BLL above 20 μg/dL. Those children were visited by a multidisciplinary team who studied their residential environment, possible sources of lead exposure, and socioeconomic and housing conditions. Educational and preventive initiatives concerning hygienic and dietary habits were also undertaken within the community. The children being studied either lived in children's homes or were getting medical assistance in a Paediatric Hospital. In the Mañay et al. (2003) paper, blood lead analyses of 2351 children (aged 2–15 yr) and 45 pregnant women were discussed. Drinking water, soil, and air samples were also analyzed.

The preliminary results showed 61% of tested children had BLL above 10 μg/dL, which is considered as the medical and/or environmental intervention blood lead limit (CDC 1991). Higher BLL were found in younger children (>4 yr). Of the 45 pregnant women tested, 67% had BLL below 10 μg/dL. BLLs were significantly higher in boys (12.1 μg/dL) than girls (11.2 μg/dL) and tended to decrease with

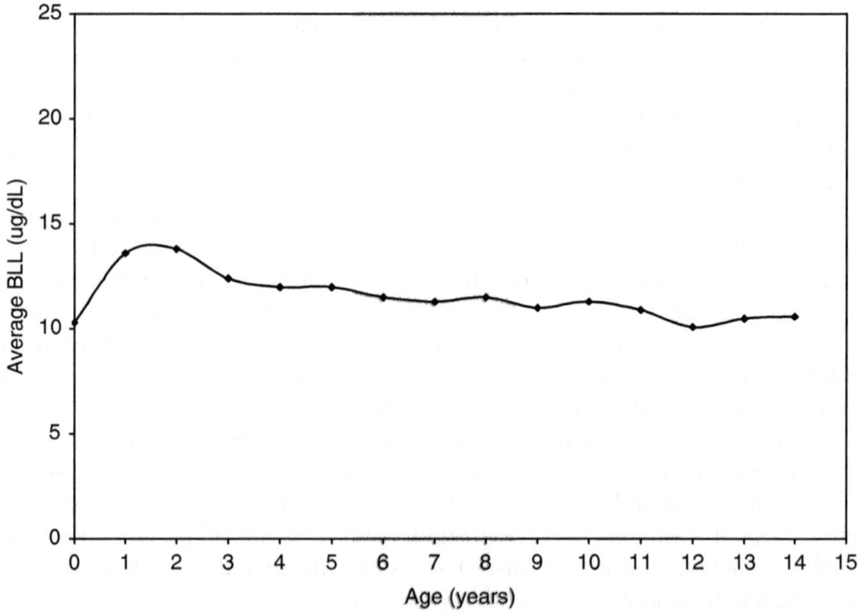

Fig. 1 Average blood lead levels (BLL) by age of children from "La Teja," Montevideo, Uruguay, 2001. (Adapted from Mañay et al. 2003, with kind permission from Salud Publica de Mexico.)

increasing age. Maximum BLLs were found in children about 2 yr old (Fig. 1), which is consistent with results of other studies (ATSDR 2003; IPCS 1995). Soil was apparently the principal source of the children's exposure because the area where they lived had been widely contaminated with lead scrap in the previous decades. This study also confirmed that the bare soil, which constituted the floor of most houses, was the main source responsible for lead uptake by children. To mitigate the health problems of those affected, priority criteria were developed to achieve favorable outcomes (reductions in BLL). These criteria emphasized education, hygiene, and nutrition.

Alvarez et al. (2003a) reported that, during 2001 and 2002, the Toxicology Department of the Faculty of Chemistry performed 10,131 lead analyses on blood samples as a community and advisory service. These samples represented 5,848 children and 1,268 adults; 3,015 samples were not specified as to gender. The data from these analyses are presented in Table 1. The average BLL found for children was 12.3 µg/dL; 60% of the values exceeded the 10 µg/dL limit (CDC 1991).

Beginning in 2001, children with BLL values higher than 20 µg/dL received public health and/or medical assistance at the Chemical Pollutants Medical Care Center of the Pereira Rossell Hospital. Those children received medical intervention, including iron supplementation and improved nutrition, and their mothers received hygienic-dietary instruction on how to reduce lead absorption. Alvarez et al. (2003b) reported follow-up testing on 387 of these children. The results of this testing statistically demonstrated that BLLs were significantly decreased with the adopted medical interventions (Fig. 2).

The Municipality of Montevideo carried out environmental surveillance of lead content in soil samples from several slum settlements associated with the environmental interventions (IMM 2003). The authors of this surveillance report attempted to assess the possible sources of lead pollution that caused the event. They suggested that the primary sources of pollution were poorly run smelters or metallurgical enterprises, deposition of industrial wastes in landfills, used battery recycling, other forms of lead waste mismanagement, burning wire for copper recovery, and contributions from vehicular traffic emissions. Soil is thus the major source of human exposure to lead. Lead-based paint and lead residues in water from lead pipelines were also recognized as potential sources of lead exposure.

In the same report (IMM 2003), the Municipality of Montevideo described results from surveillance of soil samples taken in different areas to correlate with

Table 1 Average blood lead levels (BLL) in three Uruguyan populations compared with the recommended value

	Children	Nonexposed adults	Exposed adults
Pb in blood (µg/dL)	12.3	14.6	37.1
Recommended value (µg/dL)	10	25	30
Percentage above the recommended value	>60%	13%	9%

Source: Adapted from Alvarez et al. (2003a).

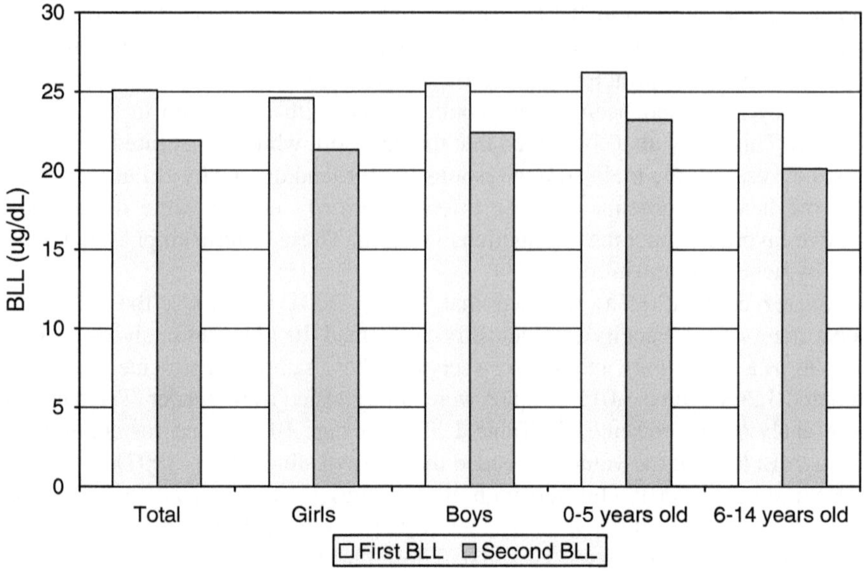

Fig. 2 Average values for first (original BLL sampling) vs. second (after medical intervention) BLL by sex and age (n = 387). (Adapted from Alvarez et al. 2003b.)

traffic lead emissions. They showed a correlation of lead levels in soil with traffic intensity in areas all over Montevideo before leaded gasoline was phased out.

2.3 *Occupational Lead Exposure*

Areas near the La Teja neighborhood were never systematically evaluated for environmental lead residues. However, when lead handling industries were operating, some studies on workers' exposure to lead were conducted. As reviewed by Mañay et al. (1999) and Pereira et al. (2003), studies on exposed workers from different manufacturing industries, i.e., battery factories, foundries, wire factories, etc., showed that almost 60% of BLL analyses made were above 40 µg/dL. In addition, a pilot study with workers from a battery storage plant who were sampled for control purposes showed mean BLLs of 48.2 µg/dL (n = 60; range, 29.0–80.0); 94% of samples exceeded 30 µg/dL (Pereira et al. 1998).

The Committee of Occupational Health from the Medical Trade Union (Danatro et al. 2001) reported that, in 2001, Uruguay's adopted lead limits for workplace environments were still those of a Ministry of Health resolution of October 1982; these limits were the same as the American Conference of Governmental Industrial Hygienists (ACGIH) reference levels for workplaces in place at that time. That legal norm did not include biological limits for lead. It was not until 2004 (MSP 2004) that the Ministry of Health first established biological reference values, "safe levels," for chemicals to which workers are exposed. It was also in 2004 that a new legal framework to regulate the control of workers' blood lead exposure levels (Legislative Power 2004a) was created. However, there is no official BLL limit for the nonoccupationally exposed adult community.

3 Description of Lead Studies and Projects in Uruguay Following the La Teja Case

3.1 Official Actions, Projects, and Reports

3.1.1 Lead in Soil

Since 2001, the Laboratory of Environmental Hygiene of the Montevideo Municipality, in coordination with the Soil Division of the Ministerial Division of Environment (IMM 2003), has worked on the study and assessment of lead-, chromium-, and cadmium-contaminated sites. Because Uruguay lacks legal norms to regulate the concentration of contaminants in soil, guideline values were adopted from international organizations (Table 2).

The results of completed studies emphasize the influence of leaded gasoline as a source of lead contamination in soil in areas of high vehicle traffic (IMM 2003). Such contamination was clearly demonstrated by comparison of lead concentrations in soil samples from urban high versus suburban and rural moderate vehicle traffic areas (Table 3).

Table 2 International recommendations for lead levels in soil

	Recommended value (mg/kg soil)	
USA (EPA)	400	Residential
Canada (Canadian Council of Ministries of the Environment)	140	Recreational

Table 3 Influence of vehicular traffic intensity on soil lead levels

Source of soil samples	Urban area	Suburban area	Rural area
Number of samples	47	16	15
Samples with lead concentration ≥400 mg/kg (%)	83	100	100
Samples with lead concentration between 140 and400 mg/kg (%)	17	0	0
Samples with lead concentration >400 mg/kg (%)	0	0	0
Average lead concentration (mg/kg)	95	33	28

Source: From IMM (2003).

Results of soil sample surveillance from various slum areas showed that, of 354 analyzed samples taken from 57 settlements, only 65 samples had residue levels above those recommended (140 mg/kg; CCME 2006), as shown in Table 4 and Fig. 3. This study was performed on samples collected by the Ministry of Health during the course of another study in which families living in their homes were being evaluated for effects of lead exposure on health. Some samples other than soil (blood) were also collected and analyzed.

3.1.2 Phase-Out of Leaded Gasoline

By 2003, a United Nations report on "Progress on Phasing out Lead in Gasoline" stated that decision makers in an increasing number of countries, including Uruguay, had recognized that eliminating the use of lead additives is a cost-effective way of reducing lead poisoning, especially in children; as a result, the phase-out of lead from gasoline had gained wide support. The phase-out of leaded gasoline in Uruguay was gradual, as described in reports from the United Nations Environment

Table 4 Distribution of lead levels in soils from settlements studied by the Laboratory of Environmental Hygiene

Number of settlements	57
Number of samples	354
Number of samples with Pb < 140 mg/kg	209
Number of samples with Pb between 140 and 400 mg/kg	80
Number of samples with Pb in soil > 400 mg/kg	65
Number of settlements having at least one sample with Pb > 400 mg/kg	19

Source: From IMM (2003).

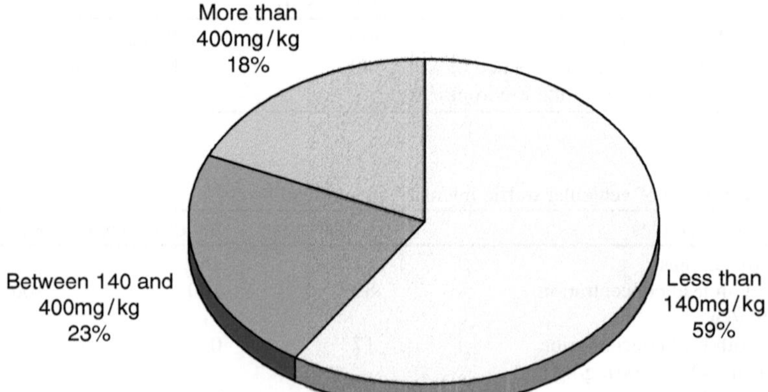

Fig. 3 Lead levels in soil from settlements studied by the Laboratory of Environmental Hygiene (IMM 2003)

Program (UNEP/OECD 1999, UNEP 2002). Finally, in November 2003, the State Refinery (ANCAP) began production of lead-free gasoline, which was first marketed in January 2004 (UNEP 2007).

3.1.3 Industrial Surveillance

In 2001, more than 100 enterprises, affiliated in some manner with lead contamination, were targeted for soil sampling by the Laboratory of Environmental Hygiene of the Montevideo Municipality. A total of 427 soil samples from industrial premises and their surroundings were taken. Resultant lead levels from these samples, as analyzed by atomic absorption spectrometry, ranged from nondetectable up to "hundreds of grams per kilogram of soil" (IMM 2003).

The variability of results was concluded to result from the following:

- Differences among the surveyed industries in the way they created environmental emissions, the degree to which their handling of wastes was inappropriate, etc.
- Distance from sampling site to primary site of lead production or lead waste processing (levels generally diminished with increasing distance).
- Duration of poor lead handling practices. Because lead is a permanent contaminant and can be mobilized by wind, rain, etc., high lead levels can persist at facility sites and their surroundings long after a company ceases to operate.

3.1.4 Lead Remediation

The Montevideo Municipality established an agreement with the Faculty of Agriculture of the State University in June 2002 to study the possibility of remediating lead-contaminated urban areas. The remediation technique consisted of applying phosphorus from apatite mineral, a natural compound of calcium phosphate, to the contaminated soil. Apatite (with lead) forms a mineral called piromorphite. The resulting mineral is less bioavailable because it is an insoluble lead compound that is not easily absorbed by humans.

In 2002, major work was undertaken to diagnose the degree of contamination at selected sites in the La Teja neighborhood. Remediation and follow-up took place during 2003 (IMM 2003).The final report of the Faculty of Agriculture was scheduled for delivery late in 2004 but is not yet available.

3.1.5 Groundwater

In September 2003, the Montevideo Municipality, along with the National Mining and Geology Direction of the Ministry of Industry and Energy, agreed to undertake a groundwater study in lead-contaminated areas. Pilot studies were conducted in various neighborhoods to better select areas deserving future attention (IMM 2003). Data from these undertakings are expected in the future.

3.1.6 Other Environmental Projects and Current Action

After 2001, several areas with different landfills of industrial origin (not involving lead scrap) were discovered and were considered a potential health risk because they potentially could turn into slum settlements. The Ministry of Housing and Environment took official action to minimize health and environmental risks in such polluted areas beginning in 2001. Environmental authorities, with help from international sources, established criteria to assess different environmental issues and pollution sites of concerns. Unfortunately, few data from these actions have been published, other than those concerning new environmental lead regulations, as discussed next.

In 2003, an inventory of contaminated sites in a southwest industrial area of Uruguay was assembled by the Ministry of Housing and Environment. The inventoried region once had both lead smelters (recently closed) and a lead-acid battery factory. These sites were environmentally profiled regarding potential exposure risks (DINAMA 2004). The conclusion was that, although there were some cases of negative impact on the environment, they did not demand urgent attention. No further data on this inventory are yet available.

3.2 Research Studies

3.2.1 Children

Several research studies addressed the health status of Uruguayan children in the context of the La Teja neighborhood case.

A study on lead-exposed children was evaluated by Cousillas et al. (2003). The studied population comprised 333 children whose BLLs were analyzed before 2001. These children either lived in polluted areas surrounding smelters, or near high vehicular traffic areas where leaded gasoline was used, or had families working with lead at home. The average age of children studied was 5.9 yr. These children had average BLL values of 15.7 μg/dL; by comparison, 60% of children from all lead-polluted areas studied had BLL values that exceeded the intervention level of 10 μg/dL (CDC 1991).

Cousillas et al. (2005) conducted field studies on randomly sampled volunteer children (0–14 yr) from Montevideo and one rural area of Uruguay. These children were divided into three groups:

- The first group comprised children ($n = 112$) who were presumably only exposed to ambient environmental pollution; these were regarded as a control group. This group represented three residential areas of Montevideo (107 children) and one rural location (5 children) (Table 5).

Table 5 Distribution by age and gender of children exposed only to ambient levels of environmental lead

Center	Children	Number	Age (yr)
Centers I, II, III[a]	Total	107	7.8 (2–14)
	Girls	51	7.6 (3–14)
	Boys	56	7.5 (2–14)
Rural center			
	Total	5	8.0 (6–11)
	Girls	3	8.6 (7–11)
	Boys	2	7.5 (6–9)

[a] Centers I–III represent three residential areas of Montevideo.
Source: Reproduced from Cousillas et al. (2005), with kind permission from Springer Science + Business Media

Table 6 Distribution by age and gender of children known to have had lead exposure

Children	Number	Age (yr)
Total	62	7.5 (0–14)
Girls	25	6.8 (0–14)
Boys	37	7.5 (0–14)

Source: Reproduced from Cousillas et al. (2005), with kind permission from Springer Science + Business Media

- The second group was composed of children (n = 62) who lived in an area known to be contaminated from the operation of an iron and lead scrap smelter (Table 6).
- The third group comprised four siblings (aged 4–13 yr) whose father recycled batteries at home.

Data were collected on each individual, with particular attention paid to age, where they lived and attended school, intensity of traffic near their homes, and smoking habits of their parents.

In the first group (ambient environment), average BLL was 9.4 µg/dL. This level is definitely higher than those reported for correspondingly exposed children in other countries. Approximately 30% of the tested children presented with BLL values above the CDC (1991) intervention level (10 µg/dL). Children from the first group who resided in the rural location had low BLL values, but those BLLs showed an inverse relationship with age (Fig. 4). No significant correlation between BLL and sex was found in this population of Uruguayan children.

Obviously, lead exposure is higher in areas where lead-related industries contributed to the environmental pollution of air, soil, and water. All 62 children in the second group had high BLLs. Soil samples from the second group (residential areas) were analyzed and found to be contaminated; soil lead levels varied from 0.1 to 2.1 mg/g (Schutz et al. 1997). Average BLL in the 62 children from the second

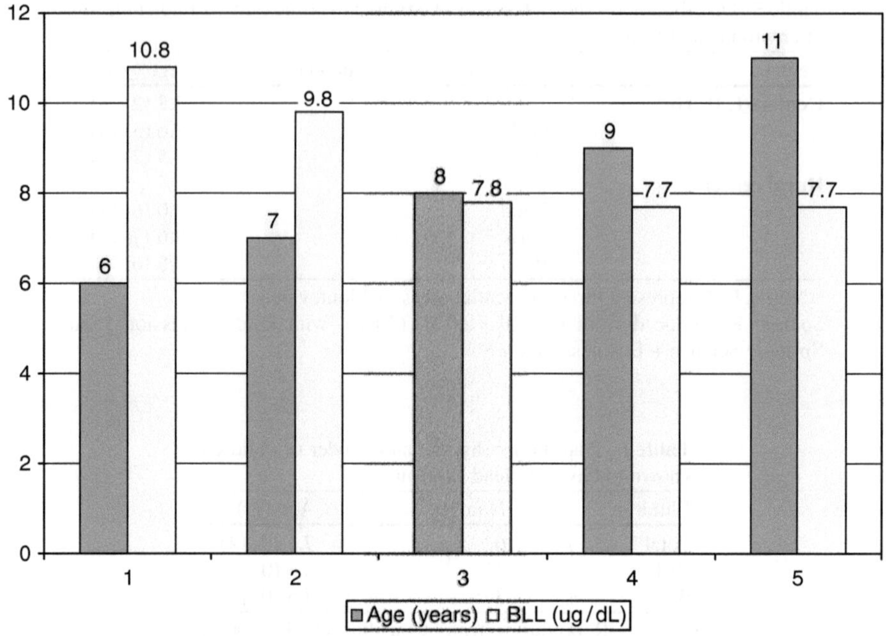

Fig. 4 BLL by age from rural children exposed to lead in the ambient environment. (Adapted from Cousillas et al. 2005, with kind permission from Springer Science + Business Media.)

group was 11.8 μg/dL, with 59% of the children having values above 10 μg/dL and 29% above the WHO Limit Value (15 μg/dL) (OMS 1980).

Results from the third group of children (four siblings) are shown in Fig. 5. All these children had had very high exposure. Lead residues were found in the soil of their garden (86 mg/g) and in window dust samples taken from their home (41 mg/g). Lead body burden was studied (CDC 1991) for these four exposed siblings to determine if treatment with chelates would be appropriate. Such treatment with calcium ethylenediaminetetraacetic acid (EDTA) can dramatically reduce BLL and bone burden. To judge treatment effectiveness, the amount of lead excreted in urine (μg) is monitored in relation to the dose of EDTA administered (mg). The test is positive if the ratio of excretion is greater than 0.6 (calculated as μg Pb excreted in urine/mg Ca EDTA dose). Results were 0.3 (4 yr old), 0.6 (7 yr old), 0.7 (11 yr old), and 0.5 (13 yr old); therefore, the physicians decided not to give the treatment.

Preliminary results comparing BLLs from two similar populations of children, one monitored for exposure in 1994 and the second in 2004, were obtained as described by Alvarez et al. (2005) and Cousillas et al. (2004). BLLs of the 60 children studied (average age, 5.2 yr) in 1994 and of 180 children (average age, 6.3 yr) studied in 2004 were compared to observe differences. Both populations were sampled at the same health care center under similar conditions. Table 7 shows the results obtained for both populations. Results indicated that the children's BLL values diminished in a statistically significant way between 1994 and 2004 ($P < 0.0001$).

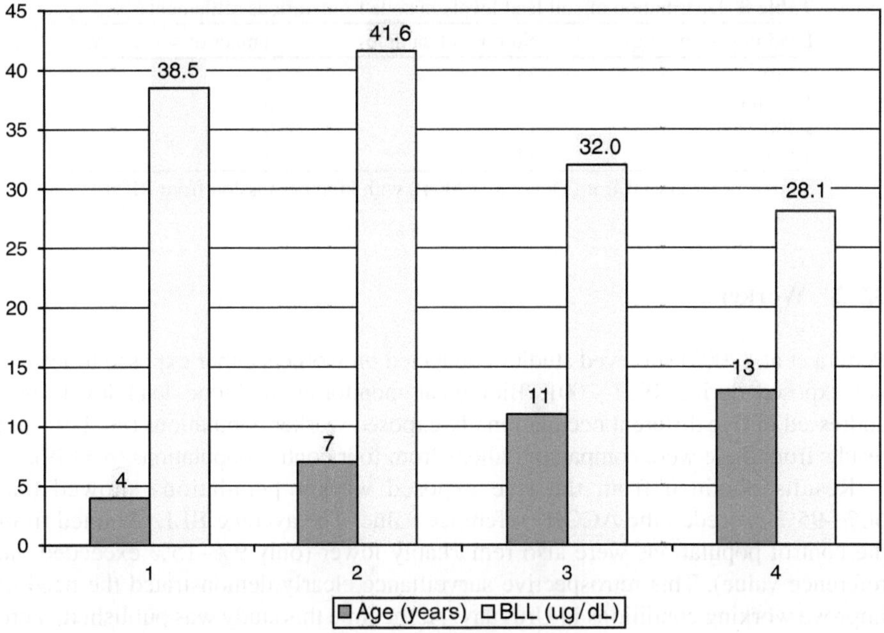

Fig. 5 BLLs by age in four children exposed to lead in their homes. (Adapted from Cousillas et al. 2005, with kind permission from Springer Science + Business Media.)

Table 7 Comparison of BLL between populations of children sampled 10 yr apart

Year	*n*	Average BLL (μg/dL)	% BLL > 10 μg/dL
1994	60	9.9	41.7
2004	180	5.7	6.7

Source: Adapted from Alvarez et al. (2005).

Possible reasons for the BLL decrease include the following:

- After rising concern for lead pollution, beginning in 2001, people became sensitive to the possible effects of lead pollution on children's health and took action to make changes in nutrition and hygiene habits.
- The phase-out of leaded gasoline.
- Action was taken to replace lead water pipes with plastic pipes in many households.

Dol et al. (2004) reported analyses of 44 samples from 15 slum settlements; 19 samples had soil lead levels higher than those recommended for housing by the Canadian guidelines (CCME 2006) (Table 8). The BLL values found in children were directly proportional to the lead level values found in soil (Dol et al. 2004). As part of this study, the Ministry of Health also ordered BLL analysis in children under 15 yr of age when the soil lead levels were found to exceed international limits.

Table 8 Distribution of soil lead levels in soils from various settlements

Lead in soil (mg/kg)	Number of samples	Number of settlements
<140	25	4
140–400	7	3
>400	12	8
Total	44	15

Source: Reproduced from Dol et al. (2004), with kind permission from RETEL.

3.2.2 Workers

Pereira et al. (2003) surveyed studies conducted on workers either exposed to lead or not exposed during 1991–2001. Biological monitoring of blood lead levels was addressed in five different occupationally exposed worker populations ($n=219$), and results from these were compared to those from four control populations ($n=139$).

Results obtained from the five exposed worker populations showed that 60%–95% exceeded the ACGIH reference value. The average BLL obtained from the control populations were also remarkably lower (only 9%–15% exceeded the reference value). This retrospective surveillance clearly demonstrated the need to improve working conditions in Uruguay. At the time this study was published, there was neither official action being taken to redress lead exposure nor specific occupational safety and health assessments being made to address the concern for lead pollution in Uruguay.

3.2.3 Nonoccupationally Exposed Adults

As a consequence of the attention paid to lead pollution after 2001, a need existed for a systematic study of BLLs in the Uruguayan adult population. In particular, reference background data on "unexposed" (those exposed only to background or ambient environmental concentrations of lead) Uruguayans for future research purposes became evident. The Toxicology and Environmental Hygiene Department, together with a private occupational health clinic, undertook an epidemiological study to address the need (Barañano et al. 2005). The research aims were to determine BLLs of working, but nonoccupationally exposed (to lead), adults living in selected areas of Montevideo. Gasoline sampling was also undertaken to measure lead content before and after the phase-out of leaded gasoline (December 2003).

A population of 700 volunteers between 20 and 64 yr who lived in Montevideo was selected for blood lead analysis and monitoring of other health parameters. The city was divided into five geographic sections according to population density, traffic intensity, industrial areas known for handling metals, and meteorological factors such as wind intensity. The results showed an overall average BLL of 5.46 µg/dL (range, 2.0–21.5 µg/dL). Considering that the international reference value for adults in the general population is a BLL less than 25 µg/dL (WHO 1995), this study reveals that these background levels for Montevideo adults are lower than the reference level and also very far below the reference limit of ACGIH for lead-exposed workers (BLL <30 µg/dL).

3.2.3 Dogs as Sentinels of Lead Pollution

Uruguay still lacks a relevant surveillance screening program for lead-polluted areas. In studying children's risk of lead exposure, Mañay et al. (2005) discovered that dogs can be very useful as sentinels for environmental lead pollution. Dogs may show early symptoms of lead intoxication at statistically significant lower BLLs than those found in children. Moreover, dogs can also have significantly higher BLL than those in children when exposed to and living in the same polluted area.

Two populations of dogs were studied: a preliminary group of stray dogs ($n = 48$), and a main group comprising randomized stray dogs ($n = 49$) and pet dogs ($n = 151$). BLLs were determined for all animals to evaluate the significance of variables such as age, sex, size, area of residence, and possible sources of lead and lead-related symptoms. Associations between BLL and single variables were then evaluated using statistical analysis.

Simultaneously, a BLL surveillance screening study on Uruguayan children was underway for an advisory report to the Municipality as described by Cousillas et al. (2003) and Mañay et al. (2005). This screening study was also made up of two groups, a preliminary one ($n = 34$); and a main group ($n = 134$), both randomly selected in the same way as were dogs. This similarity allowed the authors to compare BLL in dogs with those in children. Both studies were performed independently with comparable parameters and also included internal and external quality controls for the analytical methods and data processing.

The results (Table 9) showed higher BLL for dogs (mean, 16.3 µg/dL) than for children (mean, 9.7 µg/dL). The number of dogs with BLL <greater than> 10 µg/dL was 40% greater than for children. Significant statistical relationships between

Table 9 BLL in comparable studies with dogs and children: (I) preliminary study; (II) main study

Dogs	*n*	Mean (µg/dL)	BLL> 10 µg/dL (%)
Stray dogs (I)	48	14.6	81
Stray dogs (II)	49	16.3	84
Pet dogs (II)	151	16.0	56
Children	***n***	**Mean (µg/dL)**	**BLL> 10 µg/dL (%)**
Total (I)	34	9.5	59
Total (II)	134	9.7	35
Girls (II)	38	9.4	32
Boys (II)	96	9.8	36
<6 years of age	37	10.8	69
≥6 years of age	33	9.0	28

Source: Adapted from Mañay et al. (2005).

BLL in dogs were discerned for the following variables: age ($P < 0.001$), size ($P < 0.0001$), area of residence ($P < 0.01$), and lead-related symptoms ($P < 0.0001$). Similar statistical correlations for age, sex, area of residence, traffic intensity, and parental smoking habits were also observed for children. Correlations in BLL between pet dogs and children were tested in 12 La Teja neighborhood homes inhabited by lead-exposed families. BLLs in pet dogs were significantly higher than those for children living in the same family ($P < 0.01$) (Fig. 6).

It was concluded that a systematic surveillance on BLL in dogs can be a very useful tool to prevent and assess lead risk for children when alternatives are not available. This procedure has several advantages: lower cost than some environment screening programs such as biomonitoring (children) or analyzing environmental samples (e.g., soil, air and water); the greater sensitivity to lead of dogs; and the technique is an *in vivo* method that directly correlates to lead bioactivity. The use of dogs as sentinels may also be useful to health and environment authorities in developing countries as a first step to diagnose lead pollution problems.

3.2.4 Changes of BLL in Uruguayan Populations

Lead exposure risks have been a matter of public concern in Uruguay since 2001 when lead pollution first received official attention. In response to the public con-

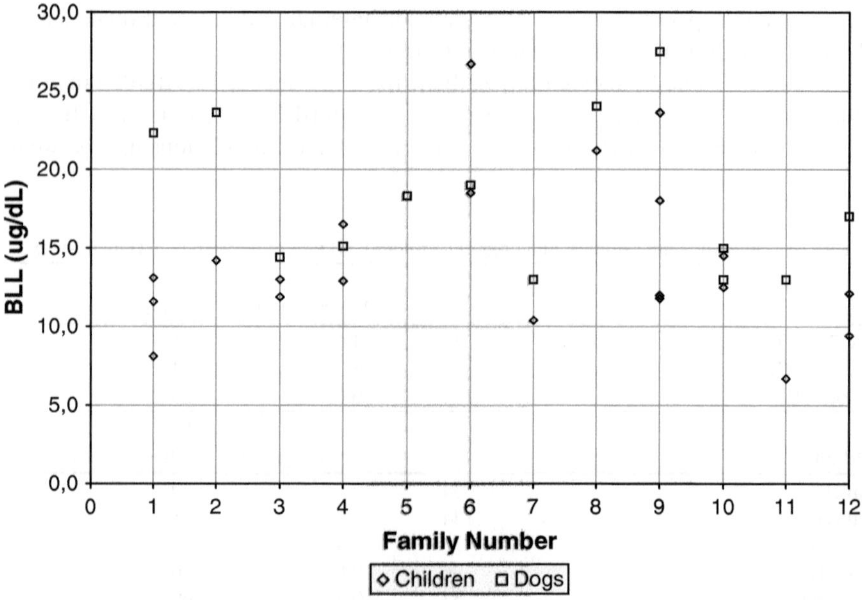

Fig. 6 BLLs in pet dogs and children from 12 "La Teja" neighborhood families. (Adapted from Mañay et al. 2005.)

cern, social and political action was taken and included creation of new regulations designed to provide for health risk management and control.

Consistent with actions to reduce public exposure to lead, Mañay et al. (2006) reviewed several studies and compared changes in BLLs in Uruguayan populations over a 10-yr period. This report included comparisons of similar populations (children, exposed and unexposed adults, and lead workers) that were sampled within this 10-yr period. BLL determinations were performed by the toxicology team at the Faculty of Chemistry using appropriate quality control methodology.

The study, which was carried out in 2004, involved sampling three populations: children (n = 180), nonoccupationally exposed adults (n = 714), and lead workers (n = 81). A framework was established to correlate BLL with variables such as age, sex, area of residence, available environmental lead data, and possible lead exposure sources (Table 10). To assess the change in risk, analytical results were statistically compared with similar screening study results performed in 1994. Blood samples were analyzed using atomic absorption spectrometry with appropriate quality controls. Results showed significantly lower BLL levels in children (5.7 μg/dL) and nonoccupationally exposed adults (5.5 μg/dL) than similar populations sampled in 1994 (9.9 μg/dL and 9.1 μg/dL, respectively). Children in 1994 showed a positive relationship between BLL and traffic intensity, and 40% of their BLL exceeded 10 μg/dL, while in 2004, only 7% were above that intervention level. These changes suggest a decrease is occurring in the contribution of environmental lead to the overall exposure of children and nonoccupationally exposed adults in Uruguay.

Workers occupationally exposed to lead did not show significant BLL differences between 1994 and 2004, and they showed mean values (49.0 Pb μg/dL vs. 42.0 Pb μg/dL, respectively) that exceeded the ACGIH Biological Exposure Index (BEI) of 30 μg/dL. It is concluded, from this comparative study, that although Uruguay still lacks a surveillance screening program for lead-polluted areas, significant improvements in preventing nonoccupational lead exposure have been made. This outcome can presumably be attributed to the phase-out of leaded gasoline and to improvements in nutrition, hygiene, and related habits for children as well as a favorable response to the official multidisciplinary actions taken. New laws have also been approved to address lead occupational exposure that require lead content of blood to be periodically monitored as part of the workers health certificate protocol.

Table 10 BLL from different Uruguayan human populations sampled in 2004 versus reference limit values and incidence above those limits

Populations (2004)	n	Average BLL (μg/dL)	Range of BLL (μg/dL)	Reference BLL (μg/dL)	BBL> reference (%)
Children	180	5.7	3.0–16.0	>10	6.7%
Nonoccupationally exposed adults	714	5.5	3.0–24.0	>25	0%
Occupationally exposed adults	81	41.9	9.0–69.0	>30	76.5%

Source: Reproduced from Mañay et al. (2006), with kind permission from Metal Ions in Biology and Medicine, vol 9

4 Legal Framework

At present, lead pollution issues in Uruguay are managed under several new regulations.

4.1.1 Law N° 17.774

"Lead in Blood Analysis in Exposed Workers" (Legislative Power 2004a) states that the determination of BLLs is mandatory and must be included in the examinations for workers health certificates. The law requires venous blood to be sampled and analyzed only by laboratories officially recognized by the Ministry of Health.

Workers with high exposure risk (foundries, lead battery factories, plumber soldering of lead, among others) must have their BLL checked twice annually. Workers with medium exposure must have BLL checked annually, whereas those with low exposure are checked every 2 yr. Office workers from lead industries may also be monitored, but at less frequent intervals than those mentioned. Another important legal change requires plumbism to appear on the list of diseases reportable by physicians to the Ministry of Health and to the State Insurance Agency.

In addition, this law addresses other procedures important to managing lead exposure. For example, there are recommendations for proper lead-bearing dust cleanup (wet instead of dry aspiration), handling of production process waste, recycling, or storage. Moreover, this law establishes some mandatory precautions, e.g., industrial effluents must be collected and evacuated in closed pipelines to prevent environmental pollution; wastes and effluents must be removed to treatment plants; and extraction devices must contain proper filters and be of a height to prevent exposure by inhalation.

There is a specific prohibition against lead solid wastes going to landfills, or being stored in buildings or other facilities, if such waste poses a risk to the public health or environment. Furthermore, new occupational rules require employers to provide separate dressing rooms for exposed workers; street and work clothes must be separated and work clothes must be washed separately.

Before eating, workers must change clothes and dining rooms must be independent from associated work areas. Employers must also provide workers with the proper clothing, shoes, gloves, or other necessary protective equipment.

4.1.2 Law N° 17.775

This law addresses other lead contamination issues: "Prevention of Lead Contamination" (Legislative Power 2004b).

- After December 31, 2004, gasoline with more than 13 mg/L lead is banned. New motor vehicles must be adapted to use unleaded fuel, and leaded gasoline must be fully removed from all dispensatories (gas stations) after December 2004.

- Lead-bearing paints cannot contain more than the maximum lead level allowed by a future ruling.
- Containers with leaded products must carry a label in Spanish, which must give the lead content and provide precautionary directions for use.
- Pipes for the distribution of water intended for human or animal use must not be made of lead, and alternative materials (i.e., plastics, galvanized iron) must not contain more than 8 % Pb; the rule also sets a maximum limit of lead in solder of 0.2%.
- Lead is banned from toys and other products used directly and frequently by children and adolescents.
- It is forbidden to store food, whether produced locally or imported, in leaded containers.
- All lead-containing products must be clearly so labeled, including the percent of lead content.
- A national register must be kept for all lead processing industries and commercial lead-containing products, including origin, storage, transit, and destination of such products.
- The law prohibits the disposal or deposit of lead-containing wastes in soil without permission. The Ministry of the Environment, together with the Municipalities, are responsible for the survey and analysis of soil in contaminated areas.
- Lead-containing storage batteries must be returned to the producers or importers after use for proper disposal; noncommercial storage batteries must be delivered to the Municipalities.

In 2002, a National Commission was created to survey, using a multidisciplinary approach, potential areas with lead (or other chemical) contamination for the purpose of taking preventive actions necessary to preserve human health. This Commission, officially at work since 2003, is composed of members from the Ministry of the Environment, Ministry of Health, Ministry of Work and Social Security, Ministry of Industry and Energy, University of the Republic, Congress of Municipalities, and Bureau of Planning and Budget, and it is led and coordinated by the Public Health Care Division of the Ministry of Health.

In 2004, the Ministry of Health issued a National Code concerning mandatory declaration of diseases associated with lead (national decree 64/004) and established a maximum permitted BLL of 15 μg/dL. The decree also included exposure risk limits for other pollutants, such as mercury, and some pesticides (Executive Power 2004). Also in 2004, a health ministerial decree (337/004) established reference values for health vigilance of chemically exposed workers. In this document, biological monitoring of BLLs is asked for; the maximum limit for BLL was set at 30 μg/dL (MSP 2004).

Another decree, number 373/2003, regulates the disposal of lead-acid batteries. It was issued by the National Direction for the Environment (DINAMA) of the Ministry of Housing and Environment (Executive Power 2003). Among its many articles, this decree requires management, recovery, and appropriate final disposition of batteries or electric storage batteries of lead-containing components and acid, to be carried out in a manner that will not damage the environment. The decree also prohibits the placement, storage, transportation, or improper processing of lead-containing batteries. The decree implements a master plan for recovery, return, and/or final destination of used batteries. At the same time, users or final consumers or holders of used batteries should return such batteries only to special reception centers capable of properly handling them. Batteries are also banned from disposal in home waste. Several examples of registration forms to gain approvals of disposal activities controlled by the National Direction for the Environment from the Ministry of Environment are also provided for by the decree.

5 Summary

Lead, ubiquitous in the environment as a result of mining and industrialization, is found as a contaminant in humans although it has no known physiological function there. Lead-exposed children are known to be the population with the highest potential health risks. The recommended biomarker to assess environmental lead exposure in animals is lead level in blood. Before 2001, the Department of Toxicology and Environmental Hygiene was the only team to produce human monitoring data on Uruguayan populations (Manay 2001a,b; Mañay et al. 1999).

Lead pollution in Uruguay first received official attention during the 2001 La Teja poisoning episode. It was in the La Teja neighbourhood of Montevideo that high BLL were found in children (as high as 20 µg/dL), prompting corrective responses from Health and Environmental authorities.

Growing awareness of environmental lead pollution and consequential human health effects from that event, resulted in public debate and demands for solutions from Health and Environmental authorities. Citizens demanded public disclosure of information concerning lead pollution and wanted action to address contaminated Uruguayan sites. In response, the Ministry of Health assembled an interinstitutional multidisciplinary committee, with delegates from health, environmental, labor, educational, and social security authorities, as well as community nongovernmental organizations (NGOs), among others. The University of the Republic was designated to serve as the main responsible entity for technical advice and support.

After 2001, new research on lead pollution was undertaken and included multidisciplinary studies with communities in response to health risk alerts. The main emphasis was placed on children exposed to environmental lead.

Major sources of Uruguayan lead contamination, similar to those in other developing countries, result from metallurgical industries, lead-acid battery processing, lead wire and pipe factories, metal foundries, metal recyclers, leaded gasoline (before December 2003), lead water pipes in old houses, and scrap and smelter solid wastes, among others. Nonoccupational lead exposure usually results from living in or near current or former manufacturing areas or improper handling of lead-containing materials or solid wastes (a particularly important health risk for children).

In this chapter, we reviewed available studies published or reported after the pollution events first announced in 2001. These studies include data on exposure, health, and actions taken to mitigate or prevent lead exposure from pollution events in Uruguay. Uruguay adopted CDC's 10 μg/dL as the reference BLL for children (CDC 1991) and a BLL of 30 μg/dL for workers (from the ACGIH standard). Environmental authorities adopted the Canadian reference concentrations for soil: residential and playgrounds (>140 mg/kg) or industrial areas (>600 mg/kg) (CCME 2006).

Most studies reviewed addressed soil pollution as the main source of lead exposure. Results of thousands of analyses indicated that most children had BLL above reference intervention limits. A significant decrease in BLL was also found over time in the study results, demonstrating the importance of medical intervention, nutrition, and environmental education. The severity of lead pollution discovered required official governmental actions, both to reduce sources of lead contamination and to address the health implications for children who had been exposed to environmental or industrial lead pollution.

Dogs were discovered to be useful sentinels for environmental lead pollution; they had higher BLL than children when exposed to the same polluted environment and developed symptoms of lead intoxication earlier and at lower BLL than did children. This same pattern was also observed in families with children and pet dogs living in the La Teja neighborhood. This discovery renders dogs prospectively useful in lead pollution monitoring and diagnosis, particularly in developing countries.

BLL results from similar human lead exposure studies conducted 10 yr apart showed significant BLL reductions, after 10 yr, for nonoccupationally exposed Uruguayans. The phase-out of leaded gasoline is thought to have contributed to this improvement.

New laws to address occupational and environmental exposures were passed to prevent new cases of lead contamination, and new research studies are underway to monitor lead pollution. Moreover, a systematic surveillance screening program for lead workers and children is planned, although it is not yet underway. The sensitization of the public to the lead pollution problem has been a key driver of governmental action to mitigate and prevent further lead pollution in Uruguay.

The changes made since 2001 appear to have yielded positive results. BLL from different populations studied more recently show decreased lead levels, suggesting a lower contribution of environmental lead to exposure of children and nonoccupationally exposed adults. The diverse analytical data collected on lead pollution in Uruguay between 2001 and 2004 were the main ingredient that allowed effective identification of lead pollution in Uruguay and paved the way for official intervention to prevent new pollution events. Nevertheless, full research studies must still be done, including both spot analysis of environmental soil, air, and water samples, and extensive screening of BLL.

Future health and environmental actions are needed, not only to remediate known areas of lead pollution, but also to investigate other sources of potential health risks.

References

ACGIH (2001) TLVs and BEIs. American Conference of Governmental Industrial Hygienists. Based on the Documentations for Threshold Limit Values for Chemical Substances and Physical Agents: Biological Exposure Indices.

ACSH (2000) American Council on Science and Health, Inc. Lead and Human Health. An Update, 2nd Ed. http://www.acsh.org-publications-boolets-lead-update.pdf.

Alvarez C, Piastra C, Cousillas A, Mañay N (2003a) Importancia del dato analítico en la contaminación por plomo durante 2001–2002 en Uruguay. XIII Congresso Brasileiro de Toxicologia. Londrina, Brasil.

Alvarez C, Piastra C, Queirolo E, Pereira A, Mañay N (2003b) Evolución de plumbemias en niños de Montevideo-Uruguay. XIII Congresso Brasileiro de Toxicología. Londrina, Brasil.

Alvarez C, Cousillas A, Pereira L, Piastra C, Heller T, Viapiana P, de Mattos B, Rampoldi O, Mañay N (2005) Estudio preliminar de niveles de plomo en sangre en niños de Uruguay. VII Congreso. SETAC, L.A. Santiago de Chile.

Amorin C (2001) Plomo para toda la vida: La verdadera historia de una contaminación masiva. Ediciones de Brecha "Nordan Comunidad."

Ascione I (2001) Intoxicación por plomo en pediatría. Arch Pediatr Urug 72(2):133–138.

ATSDR (2003) Agency for Toxic Substances and Disease Registry. The nature and extent of lead poisoning in children in the United States. A report to the Congress. U.S. Department of Health and Human Services. Atlanta, GA.

ATSDR (2005) Agency for Toxic Substances and Disease Registry. Toxicological profile for lead. Atlanta, GA. http://www.atsdr.cdc.gov/toxprofiles/tp13.pdf.

Barañano R, Mañay N, Pereira L, Cousillas A, Moratorio G, Dibarboure H (2005) Estudio Observacional: Niveles de Plombemia en población en edad laboral activa en una ciudad rioplatense. Proceedings VII Congreso Ibero Americano de Medicina del Trabajo.

Burger M, Pose D (2001) Plomo como contaminante ambiental. Ambios Cultura Ambiental (II) 6:20–22.

CCME (2006) Canadian Soil Quality Guidelines for the Protection of Environmental and Human Health. Canadian Council of Ministers of the Environment, Ottawa.

CDC (1991) United States Center of Disease Control. Preventing Lead Poisoning in Young Children. A statement by the Center of Disease Control US. Department of Health and Human Services. Atlanta, Georgia. http://wonder.cdc.gov/wonder/prevguid/p0000029/p0000029.asp.

Cousillas A, Mañay N, Pereira L, Rampoldi O (1998) Relevamiento de plumbemias en un complejo habitacional de Montevideo (Uruguay). Acta Farm Bonaer 17(4):291–96.

Cousillas A, Mañay N, Alvarez C, Pereira L, Rampoldi O (2003) Exposición ambiental al Plomo en la población infantil de Uruguay. Estudios realizados entre 1992 y 2001. International Congress of Occupational Health (ICOH 2003), Foz de Iguazú, Brasil.

Cousillas A, Pereira L, Alvarez C, Piastra C, Heller T, Viapiana P, De Mattos B, Mañay N, Rampoldi O (2004) Estudio Comparativo de Niveles de Plomo en Sangre de Niños Uruguayos: 1994–2004. Resultados Preliminares. XXIV Jornadas Interdisciplinarias de Toxicología. III Jornadas Rioplatenses de Toxicología. Buenos Aires, Argentina.

Cousillas A, Mañay N, Pereira L, Alvarez C, Coppes Z (2005) Evaluation of Lead Exposure in Uruguayan Children. Bull Environ Contam Toxicol 75:629–636.

Danatro D, Gómez F, Laborde A, López B, Perona D, Spontón F, Tomasina F, Velázquez V (2001) Contaminación por Plomo. http://www.smu.org.uy/sindicales/resoluciones/informes/plomo-0501.html.

DINAMA (2004) Estudio Piloto para la Identificación y Caracterización de Sitios en la Microregión del Rosario. Dirección Nacional de Ambiente, Ministerio de Vivienda, Ordenamiento Territorial y Ambiente. http://www.dinama.gub.uy.

Dol I, Feola G, García G, Alonzo C (2004) Contaminacion ambiental por plomo en asentamientos urbanos de Montevideo, Uruguay y su repercusion en los nivels de plomo en sangre en población infantil. RETEL http://www.sertox.com.ar/retel/default.htm.

Executive Power (2003) Decree 373/2003. http://www.presidencia.gub.uy/decretos/2003091005.htm.

Executive Power (2004) Decree 64/2004 of Events and Diseases of Mandatory Declaration. http://www.presidencia.gub.uy/decretos/2004022007.htm.

IMM (2003) Contaminación por Metales en suelo. In: Informe Ambiental de Montevideo. Intendencia Municipal de Montevideo, Documentos de Desarrollo Ambiental. http://www.montevideo.gub.uy/ambiente/documentos/infoamb03c.pdf.

IPCS (1995) Environmental Health Criteria No 165: Inorganic Lead. International Programme on Chemical Safety, Geneva, Switzerland.

Legislative Power (2004a) Law 17774 of Lead in Blood Analysis. http://www.parlamento.gub.uy/leyes/AccesoTextoLey.asp?Ley=17774&Anchor=.

Legislative Power (2004b) Law 17775 about Prevention of Lead Contamination. http://www.parlamento.gub.uy/leyes/AccesoTextoLey.asp?Ley=17775&Anchor=.

Mañay N (2001a) Antecedentes y situación actual de la contaminación por plomo en Uruguay. Taller Latinoamericano: envenenamiento de niños por plomo. Evaluación Prevención y Tratamiento. CEPIS-OPS, Lima, Perú.

Mañay N (2001b) La Contaminacion por Plomo y su Repercusion Sobre la Salud en Uruguay: Antecedentes y Situación Actual. Revista de la Asociación de Química y Farmacia del Uruguay 31:3–8.

Mañay N, Pereira L, Cousillas A (1999) Lead contamination in Uruguay. Rev Environ Contam Toxicol 159:25–39.

Mañay N, Alonzo C, Dol I (2003) Contaminación por plomo en el barrio La Teja, Montevideo-Uruguay. In: Suplemento "Experiencia Latinoamericana" Salud Publica de México 45:268–275.

Mañay N, Cousillas A, Pereira L, Alvarez C (2005) Lead Biomonitoring on Dogs as Sentinels for Risk Environment Assessment. Proceedings, XIII International Conference on Heavy Metals in the Environment, pp 221–224.

Mañay N, Alvarez C, Cousillas A, Pereira L, Baranano R, Heller T (2006) Changes in blood lead levels in Uruguayan populations. In: Alpoim MC, Morais PV, Santos MA, Cristovao A, Centeno J, Collery P (eds) Metals Ions in Biology and Medicine, vol 9. John Libbey Eurotext, Paris, pp 530–534.

Matos V (2001) El caso de la contaminación por plomo. http://www.ambiental.net/publicaciones/MatoscontaminacionPlomo.pdf.

MSP (2004) Ordenanza 337. Ministerio de Salud Pública. Ref. 001-1190/2004.

OMS (Organización Mundial de la Salud) (1980) Límites de exposicion profesional a metales pesados que se recomiendan por razones de salud. Informe Técnico No.647. Ginebra, Suiza.

Pereira L, Mañay N, Cousillas A, Korbut S, Rampoldi O, Heller T (1998) Exposición a plomo en una fábrica de baterías en Uruguay. X ALATOX Asociacion Latinoamericana de Toxicología, Congress Book, Cuba.

Pereira L, Cousillas A, Heller T, Mañay N (2003) Biological monitoring of lead exposure of Uruguayan workers during 1991–2001. International Congress of Occupational Health, Foz de Iguazú, Brasil.

Ponzo J (2002) Contaminación por Plomo en Uruguay. Informe síntesis de la situación actual. Ministerio de Salud Pública.

Schutz A, Barregard L, Sallsten G, Wilske J, Mañay N, Pereira L, Cousillas AZ (1997) Blood Lead in Uruguayan Children and Possible Sources of Exposure. Environ Res 74:17.

UNEP (2002) Progress in phasing out lead in gasoline. http://www.unep.org/GC/GC22/Document/INF%2023%20E.pdf

UNEP (2007) The Global Campaign to Eliminate Leaded Gasoline: Progress as of January 2007 http://www.unep.org/pcfv/PDF/LeadReport.pdf.

UNEP/OECD (1999) Phasing Lead out of Gasoline: An Examination of Policy Approaches in Different Countries. http://www.oecd.org/dataoecd/36/29/1937036.pdf.

WHO (1995) Environmental Health Criteria No. 165: Inorganic Lead. World Health Organization, Geneva, Switzerland.

Applications of Carboxylesterase Activity in Environmental Monitoring and Toxicity Identification Evaluations (TIEs)

Craig E. Wheelock, Bryn M. Phillips, Brian S. Anderson, Jeff L. Miller, Mike J. Miller, and Bruce D. Hammock

C.E. Wheelock (✉)
Division of Physiological Chemistry II, Department of Medical Biochemistry and Biophysics, Karolinska Institutet, Scheeles väg 2, SE-171 77, Stockholm, Sweden

B.M. Phillips
Department of Environmental Toxicology, University of California Davis, Marine Pollution Studies Laboratory, Monterey, CA 93940, USA

B.S. Anderson
Department of Environmental Toxicology, University of California Davis, Marine Pollution Studies Laboratory, Monterey, CA 93940, USA

J.L. Miller
AQUA-Science, Inc., Davis, CA 95616, USA

M.J. Miller
AQUA-Science, Inc., Davis, CA 95616, USA

B.D. Hammock
Department of Entomology, University of California Davis, Davis, CA 95616, USA

1 Introduction

The purpose of this review is to examine uses of carboxylesterase activity in environmental monitoring with a specific emphasis on pyrethroid insecticides. The chapter begins with an overview of the enzyme class, including general structure, function, catalytic mechanism, and substrate specificity. This section serves to introduce carboxylesterases, their biological significance, and their role in metabolism and detoxification reactions. Following this section, an in-depth analysis of different reports of applications of carboxylesterase activity in environmental monitoring is presented on an organism-specific basis. From an environmental standpoint, one of the most important carboxylesterase-mediated reactions is the hydrolysis and subsequent detoxification of pyrethroid insecticides. This reaction is one of the main detoxification pathways for pyrethroids in numerous organisms ranging from worms to fish to humans and is also an important pathway for the development of insect resistance to pyrethroid-associated toxicity. Accordingly, this class of insecticide is reviewed in more detail, with emphasis on toxicity and physical properties. The high hydrophobicity of pyrethroids is specifically addressed with a discussion of the effects of surface adsorption upon the observed toxicity in aquatic testing systems. A particular point is that changing agricultural practices combined with new legislation are causing a shift in insecticide usage patterns from organophosphates (OPs) and carbamates to pyrethroids. The effects of this shift are complex and potentially far reaching, especially the environmental consequences. In particular, the extreme toxicity of pyrethroids to many aquatic organisms, combined with their hydrophobicity, has resulted in concern regarding their potential environmental effects. This concern is exacerbated by the fact that current Toxicity Identification Evaluation (TIE) protocols devised for the identification of insecticides (and other environmental contaminants) in aqueous and sediment samples do not identify pyrethroid-associated toxicity with complete certainty. To address this shortfall, the use of carboxylesterase activity to hydrolyze pyrethroids in aquatic toxicity testing has been proposed as a simple, mechanistically based method to selectively identify pyrethroid-associated toxicity. This chapter reviews TIE protocols and the role of carboxylesterase activity in the development of TIE methods. A series of case studies are presented in which carboxylesterase activity was employed to identify pyrethroid-associated toxicity. Additional methods for the selective detection of pyrethroid-associated toxicity are also examined, including the use of temperature differentials and piperonyl butoxide (PBO). The strengths and weaknesses of the carboxylesterase-addition technique are also analyzed, with a number of distinct recommendations made for future development. Taken together, this review provides a detailed analysis of multiple applications of carboxylesterase to environmental monitoring and strongly advocates for further work on this enzyme system.

2 Carboxylesterases

Carboxylesterases are enzymes in the α/β-hydrolase fold family that catalyze the hydrolysis of carboxyl esters via the addition of water, as shown in Fig.1 (Junge and Krisch 1975; Myers et al. 1988; Ollis et al. 1992; Aldridge 1993; Cygler et al. 1993; Heikinheimo et al. 1999; Oakeshott et al. 1999; Satoh and Hosokawa 2006; Hosokawa et al. 2007). The α/β-hydrolase fold is a superfamily of enzymes that also includes cholinesterases (Quinn 1987, 1997, 1999), epoxide hydrolases (Morisseau and Hammock 2005; Newman et al. 2005), and phosphotriesterases (such as paraoxonase) (Sogorb et al. 2004) as well as other enzymes (Ollis et al. 1992; Hotelier et al. 2004). In standard esterase nomenclature, carboxylesterases are termed B-esterases in that they are inhibited by OPs, as opposed to A-esterases, which are defined as hydrolyzing uncharged esters that are not inhibited by OPs or other acylating inhibitors (Aldridge 1953a,b, 1993). Carboxylesterases are found in many tissues including liver, lung, small intestine, heart, kidney, muscle, brain, testis, adipose tissue, nasal and respiratory tissues, leukocytes, and the blood (see Satoh and Hosokawa 1998, and references therein). However, carboxylesterase expression and activity are tissue- and organism dependent, with levels and activities varying widely (Imai 2006). Carboxylesterases consist of multiple isozymes that vary with both the tissue and organism (Hosokawa et al. 1995; Satoh and Hosokawa 1995; Imai 2006), making individual nomenclature complicated. These

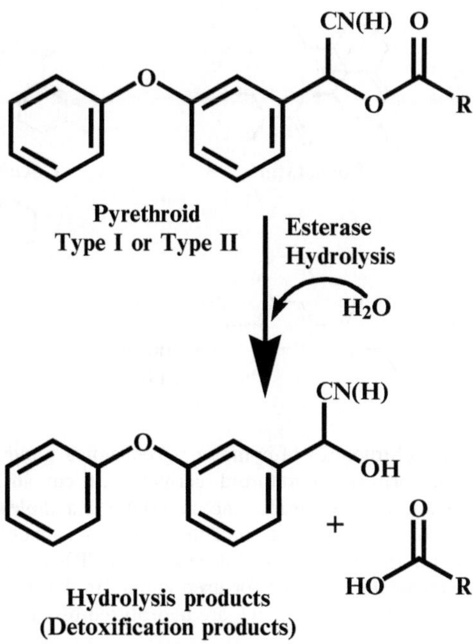

Fig. 1 Esterase-mediated hydrolysis of pyrethroids. Esterases hydrolyze an ester via the addition of water to form the corresponding alcohol and acid, which are generally detoxification products

enzymes play a significant role in the metabolism and subsequent detoxification of many agrochemicals and pharmaceuticals (representative structures are shown in Fig. 2) (Redinbo and Potter 2005; Potter and Wadkins 2006). In particular, carboxylesterases hydrolyze pyrethroids (Abernathy and Casida 1973; Stok et al. 2004a; Wheelock et al. 2004) and bind stoichiometrically to carbamates (Gupta and Dettbarn 1993; Sogorb and Vilanova 2002) and organophosphates (Kao et al. 1985; Casida and Quistad 2004). Carboxylesterases are also important in the metabolism of a number of therapeutics (Williams 1985), including the cholesterol-lowering drug lovastatin (Tang and Kalow 1995), the antiinfluenza drug oseltamivir (Tamiflu) (Shi et al. 2006), the narcotic analgesic meperidine (Demerol) (Zhang et al. 1999), and cocaine and heroin (Pindel et al. 1997). Carboxylesterase activity is also used extensively in soft- and pro-drug design (Bodor and Buchwald 2000, 2003, 2004), as demonstrated by activation of the cancer therapeutic pro-drug CPT-11 through its conversion to SN-38 (Potter et al. 1998; Wadkins et al. 2001). Given the importance of this enzyme class in metabolizing this suite of compounds, interest in the study of their function, distribution, and selectivity is greatly increasing. It should also be

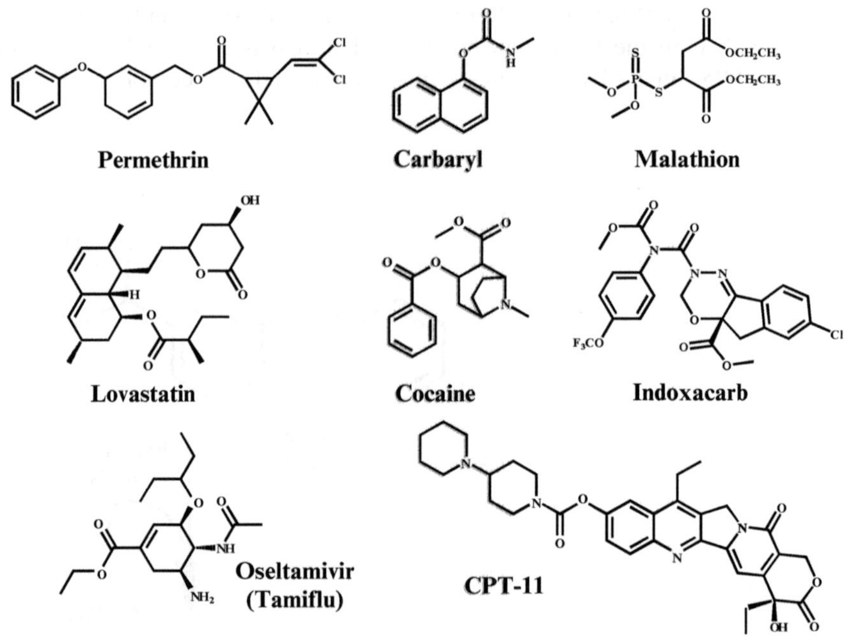

Fig. 2 Structures of common agrochemicals and pharmaceuticals that interact with esterases. Permethrin is a pyrethroid, carbaryl is a carbamate, malathion is an organophosphate (OP), indoxacarb is an oxadiazine, lovastatin is a cholesterol-lowering drug used in the treatment of cardiovascular disease, cocaine is a tropane alkaloid, oseltamivir or Tamiflu is an antiviral drug used in the treatment of influenza, and CPT-11 or Irinotecan is a chemotherapy agent that is a topoisomerase 1 inhibitor used mainly in the treatment of colon cancer. All compounds, with the exception of carbaryl, are hydrolyzed by carboxylesterases

mentioned that there are enzymes in other protein families that are not technically carboxylesterases but do hydrolyze esters; however, this review focuses solely on carboxylesterase-mediated ester hydrolysis.

2.1 *Esterase Hydrolysis: Mechanism and Inhibition*

The mechanism by which esterases hydrolyze their substrates has been examined by many research groups using both biochemical and structural means. The detailed mechanism of hydrolysis has been reviewed elsewhere (Satoh and Hosokawa 1995, 1998; Quinn 1997, 1999; Satoh et al. 2002; Sogorb and Vilanova 2002; Redinbo et al. 2003) and is only briefly presented here. Interested readers are referred to the immediately foregoing references for a more extended presentation of the hydrolysis mechanism. The publication of crystal structures of mammalian carboxylesterases (Fig. 3) has greatly contributed to our understanding of the enzyme mechanism

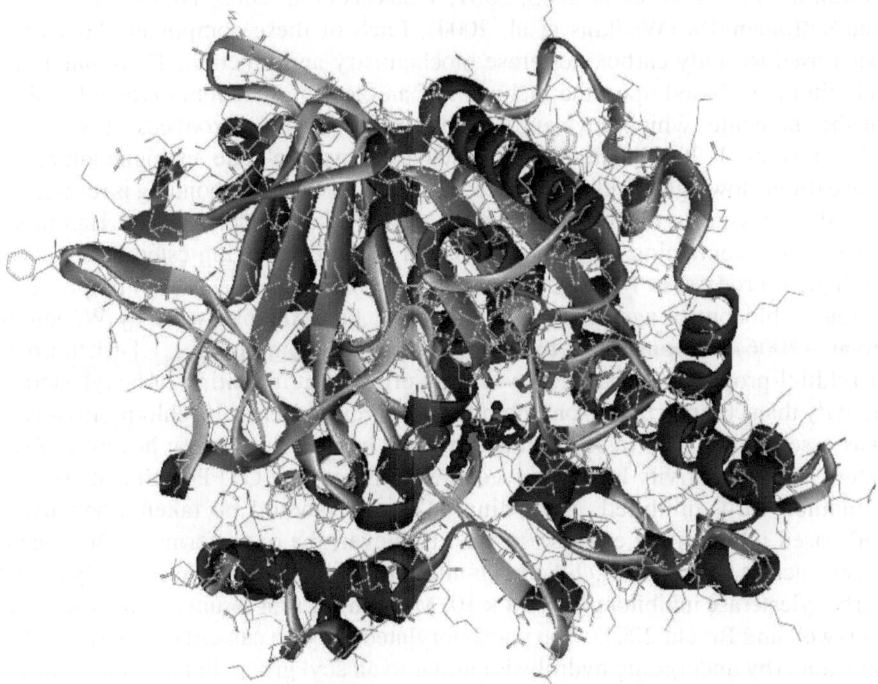

Fig. 3 Solid ribbon structure of human liver carboxylesterase 1 (hCE1) complexed with homatropine, shown as a darker ball-and-stick structure in the *lower right part* of the figure. The figure was generated from the crystal structure of Bencharit et al. (Bencharit et al. 2003a,b), from the RCSB protein data bank (http://www.rcsb.org/pdb/cgi/explore.cgi?pdbId=1MX5). [The image was created with DS ViewerPro 5.0 (Accelrys, San Diego, CA) and is reproduced from Wheelock et al. (2005c) with kind permission from the *Journal of Pesticide Science*.]

(Bencharit et al. 2002, 2003a,b, 2006; Fleming et al. 2005, 2007). The general mechanism involves a catalytic triad centered on the GXSXG active serine motif for serine esterases (Ollis et al. 1992). For carboxylesterases, this motif consists of a Ser, His, and either a Glu or Asp residue (Ser221, His468, and Glu354 for human carboxylesterase 1, hCE1) (Bencharit et al. 2003a,b); however, recent work has illustrated a potential fourth catalytic residue, which is also a serine amino acid (Stok et al. 2004b). Carboxylesterases cleave esters via a two-step process that involves the formation and degradation of an acyl-enzyme intermediate. The catalytic or nucleophilic serine is first activated to generate the oxygen nucleophile, which then attacks the carbonyl carbon of the ester substrate, leading to formation of the acyl-enzyme intermediate. The alcohol hydrolysis product is then released to undergo nucleophilic attack by water, leading to the release of the carboxylic acid and return of the catalytic amino acids to their original state.

Several different types of carboxylesterase inhibitors have been reported in the literature. The main structural motifs include trifluoromethyl ketone (TFK)-containing inhibitors (Székács et al. 1992; Wheelock et al. 2001), OP derivatives (Casida and Quistad 2004), carbamates (Gupta and Dettbarn 1993; Sogorb and Vilanova 2002), diones (Hyatt et al. 2005, 2007; Wadkins et al. 2005; Hicks et al. 2007), and sulfonamides (Wadkins et al. 2004). Each of these compound classes has been used to study carboxylesterase biochemistry and function. TFK-containing inhibitors are based upon the inclusion of an electron-deficient carbonyl moiety in the molecule, which covalently binds to the enzyme (Brodbeck et al. 1979; Wheelock et al. 2002). These compounds are transition-state analogue inhibitors that exhibit slow tight-binding kinetics; however, the covalent bonding is reversible, and the enzyme is reactivated over days to weeks (Abdel-Aal and Hammock 1986). TFK-containing inhibitors are useful tools for studying esterase-mediated biological processes (Wheelock et al. 2002, in press) and have been proposed as potential biocontrol agents (Rosa et al. 2006). For example, work by Wheelock et al. (2006) demonstrated that the TFK-based compound 1,1,1-trifluoro-3-octylthiol-propan-2-one (OTFP) was more efficient at inhibiting carboxylesterase activity than chlorpyrifos-oxon. Inhibition of porcine esterase by chlorpyrifos-oxon was essentially 100% (as determined by enzyme activity assay); however, 72 hr later, ~72% of activity had been recovered, whereas the OTFP-inhibited esterase remained fully inhibited. Accordingly, caution should be taken when using OP-based inhibitors in esterase studies. However, the oxon-forms of OP insecticides, such as paraoxon (*O,O*-diethyl *p*-nitrophenyl phosphate), are generally potent carboxylesterase inhibitors ($k_i = 1.4 \times 10^6\ M^{-1}\ min^{-1}$ for rat serum carboxylesterase; Maxwell and Brecht 2001). The phosphorylated enzyme can either release the OP substrate (by undergoing hydrolysis similar to an acyl group, but at a much slower rate), or it can undergo aging, where the enzyme is essentially catalytically dead (i.e., the OP acts as a suicide substrate) (Maxwell 1992a). A mechanistic description of these processes is shown in Fig. 4. A similar reaction can occur with carbamates; however, the methylcarbamoylated enzyme is less stable than the phosphorylated enzyme, thus accounting for the decreased toxicity of some carbamates relative to OPs (Casida and Quistad 2004). A number of papers have reviewed the

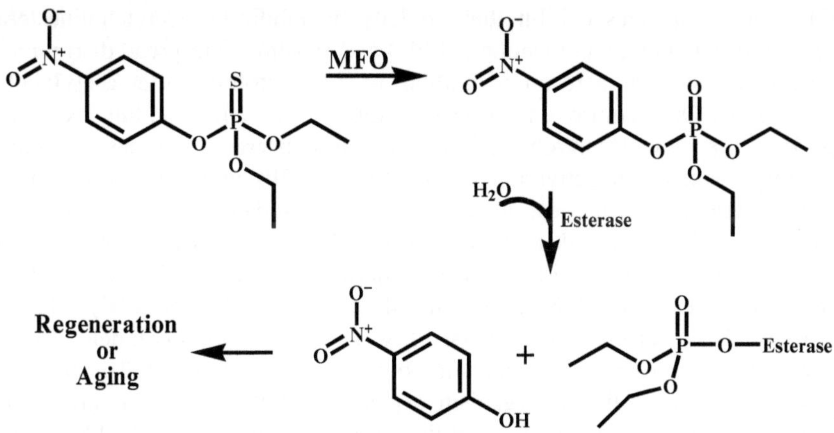

Fig. 4 Carboxylesterase inhibition mechanism for the organophosphate insecticide parathion. Parathion is first activated via mixed-function oxidases (*MFO*) to the "active" oxon form (paraoxon), which is the inhibitory structure of the compound. Paraoxon then binds to the esterase and is hydrolyzed by the addition of water, releasing *p*-nitrophenol. The phosphorylated esterase can then either release the phosphate group and regain catalytic activity, or become "aged" where the phosphate remains permanently bound and the enzyme loses catalytic activity. [Figure 4 is reproduced from Wheelock et al. (2005c), with kind permission from the *Journal of Pesticide Science*.]

interactions of carbamates and OPs with carboxylesterases (Fukuto 1990; Gupta and Dettbarn 1993; Sogorb and Vilanova 2002; Casida and Quistad 2004, 2005).

2.2 *Role of Carboxylesterases in Agriculture*

Carboxylesterases play an important role in agrochemical efficacy and detoxification (Ahmad and Forgash 1976a,b; Casida and Quistad 1998, 2004; Sogorb and Vilanova 2002; Wheelock et al. 2005c) by interacting with three major classes of agrochemicals: OPs (Kao et al. 1985; Satoh and Hosokawa 2000), carbamates (Gupta and Dettbarn 1993), and pyrethroids (Casida et al. 1983). Major pathways for agrochemical metabolism have been reviewed elsewhere, and involve a number of different enzyme systems that are beyond the scope of this review, including P450 monooxygenases (P450 MOs) (Kulkarni and Hodgson 1984), glutathione *S*-transferases (GSTs) (Fournier et al. 1992; Enayati et al. 2005), phosphotriesterases (Sogorb et al. 2004), as well as carboxylesterases (Wheelock et al. 2005c). It is well known that variability in carboxylesterase levels and relative isozyme abundance contribute to the selective toxicity of ester-containing insecticides in a range of organisms from fish to insects and mammals (Brooks 1986; Chiang and Sun 1996; Li and Fan 1997; Wheelock et al. 2003, 2005a; Huang and Ottea 2004; Stok et al. 2004a).

OPs and carbamates exhibit their toxicity by inhibiting acetylcholinesterase (Maxwell 1992b; Gupta and Dettbarn 1993) and causing widespread disruption of the nervous system as a result of buildup of the neurotransmitter acetylcholine (Quinn 1987). OPs require activation by P450s to the corresponding oxon form (Fig. 5) before they effectively inhibit acetylcholinesterase, whereas carbamates do not require metabolic activation. Carbamates and OPs bind stoichiometrically to carboxylesterases, leading several researchers to postulate that carboxylesterases act as an agrochemical sink to protect acetylcholinesterase from OP/carbamate-mediated toxicity (Gupta and Kadel 1990; Maxwell et al. 1994; Yang and Dettbarn 1998; Dettbarn et al. 1999; Sweeney and Maxwell 1999; Kuster 2005). Cleavage of pyrethroids by carboxylesterases (see Fig. 1) is one of the main detoxification pathways for this group of pesticides in both mammals (Casida 1980) and insects (Davies 1985). Accordingly, organism exposure to OPs and/or carbamates, followed by pyrethroids, can cause synergistic toxicity (Gaughan et al. 1980; Martin et al. 2003), and these effects have been demonstrated in multiple organisms (Abernathy and Casida 1973; Casida et al. 1983; Denton et al. 2003; Choi et al. 2004). For example, one of the principal metabolic pathways for malathion is via carboxylesterase-mediated hydrolysis, and the toxicity of malathion correlates with carboxylesterase levels in humans, rats, and mice (Talcott 1979; Talcott et al. 1979a,b,c). Li and Fan (1996) showed that, in five different species of freshwater fish, carboxylesterase inhibition increased malathion toxicity. Accordingly, the

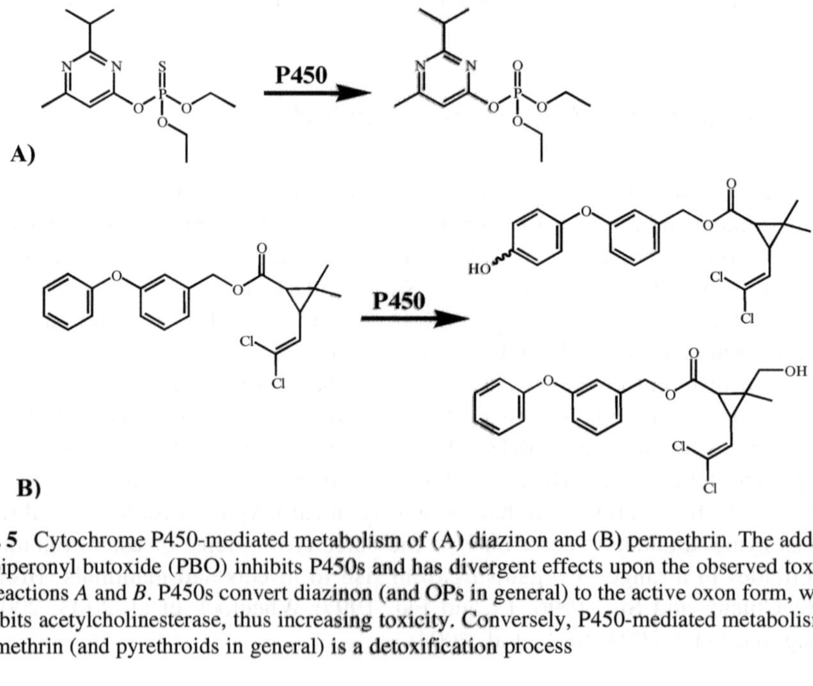

Fig. 5 Cytochrome P450-mediated metabolism of (A) diazinon and (B) permethrin. The addition of piperonyl butoxide (PBO) inhibits P450s and has divergent effects upon the observed toxicity to reactions *A* and *B*. P450s convert diazinon (and OPs in general) to the active oxon form, which inhibits acetylcholinesterase, thus increasing toxicity. Conversely, P450-mediated metabolism of permethrin (and pyrethroids in general) is a detoxification process

concept of agrochemical interactions is well established, and the importance of examining the toxicity of chemical mixtures has been emphasized in recent literature (Lydy et al. 2004).

A major concern in agrochemical use is the development of insect resistance (Casida and Quistad 1998), which can lead to increased pesticide use, cost increases, and detrimental environmental effects. A prominent pathway of resistance development to agrochemicals involves increased levels of carboxylesterases (Newcomb et al. 1997; Hemingway and Karunaratne 1998; Byrne et al. 2000; Harold and Ottea 2000; Oakeshott et al. 2005; Cui et al. 2007). This pathway has been demonstrated in multiple species of insects with OPs and pyrethroids (Immaraju et al. 1990; Elzen et al. 1992; Solomon and Fitzgerald 1993; Cahill et al. 1995; Mazzarri and Georghiou 1995; Zhao et al. 1996; Ahmad et al. 2002; McAbee et al. 2004). The study of carboxylesterases and their interactions with OPs and pyrethroids is therefore economically important. One particular area of interest is the development of isozyme-selective inhibitors of insect pyrethroid-hydrolyzing esterases. This type of inhibitor could be coapplied with pyrethroids as a synergist, in a similar fashion as PBO, to control esterase-mediated resistance. Other areas of active research include the synthesis of selective pyrethroid surrogate substrates to measure carboxylesterase activity to identify resistance mechanisms (McAbee et al. 2004; Devonshire et al. 2007). Carboxylesterase-mediated agrochemical resistance can also have ramifications for human health. Pyrethroids and OPs are extensively used for vector-borne disease control and are particularly important in the fight against malaria (Curtis and Mnzava 2000; Greenwood et al. 2005; Coleman and Hemingway 2007).

2.3 *Use of Carboxylesterase Activity in Environmental Monitoring*

Increasingly, there is need to monitor the environment for the presence of agrochemicals. Bioassays are commonly used in environmental monitoring and depend upon well-established biological endpoints that correlate with agrochemical exposure. A classic example is the use of acetylcholinesterase activity as an index of OP exposure (Fulton and Key 2001). Because exposure to OPs results in inhibition of acetylcholinesterase activity, it can be used to provide a measure of organism exposure (Rickwood and Galloway 2004). This approach has a number of advantages: (1) it is a mechanistically based biological effect that can be directly correlated with compound exposure, (2) it does not require expensive analytical instrumentation, and (3) it can be performed in the field. Positive hits can be validated, as necessary, with analytical data [i.e., gas chromatography/mass spectrometry (GC/MS) or liquid chromatography/mass spectrometry (LC/MS) data]. Although well established in the literature (Sturm et al. 2000; Fulton and Key 2001; Galloway et al. 2002; Rickwood and Galloway 2004), recent reports have questioned the use of acetylcholinesterase as a sole biomarker of OP and/or carbamate exposure (Galloway et al.

2004b; Rickwood and Galloway 2004; Wheelock et al. 2005a). Studies have shown that OPs and potentially carbamates have increased affinity for carboxylesterase over acetylcholinesterase, suggesting that carboxylesterase activity will provide a more sensitive endpoint (Gupta and Dettbarn 1993; Escartin and Porte 1997; Wogram et al. 2001; O'Neill et al. 2004; Wheelock et al. 2005a).

Because an organism's sensitivity to pyrethroid, OP, or carbamate exposure may be influenced by its endogenous carboxylesterase activity, that activity may be useful in predicting the effects of agrochemical exposure upon ecosystem health (Barata et al. 2004). However, there are currently insufficient data available in the literature to fully validate this concept. Emerging research involves the use of carboxylesterase activity as a biomarker of organism exposure or susceptibility to agrochemicals (Huang et al. 1997; Sanchez-Hernandez et al. 1998; Galloway et al. 2004b). Of the limited work performed in this area, most has focused on esterase levels in fish (James 1986; Wogram et al. 2001). Early work demonstrated that carboxylesterase levels in rainbow trout have different expression levels throughout development (Kingsbury and Masters 1972), suggesting that different life stages may vary in susceptibility to esterase inhibitors or ability to hydrolyze substrates. Further work showed that rainbow trout have lower levels of carboxylesterase activity than mammals (Glickman and Lech 1981, 1982; Glickman and Casida 1982; Glickman et al. 1982). Additionally, it has been found that activity varies among fish species (Barron et al. 1999; Wheelock et al. 2005a) and can account for differences in susceptibility to pyrethroid toxicity (Glickman et al. 1979). However the extreme susceptibility of many fish species to pyrethroid toxicity may be more a function of sensitivity of their sodium channels then low carboxylesterase activity (Glickman and Casida 1982; Glickman and Lech 1982), although carboxylesterase activity may play some role in toxicity.

Carboxylesterase activity in relation to agrochemical exposure has been examined in additional species and may therefore be useful for ecosystem-wide environmental monitoring projects (Wilson and Henderson 1992; Chevre et al. 2003; Galloway et al. 2004a,b). A significant advantage of carboxylesterase activity-based biomarkers is that assays can be nondestructive in contrast to use of brain acetylcholinesterase activity. Plasma carboxylesterase activity is generally low in mammals, but it is higher in fish, amphibian and birds (Thompson et al. 1991a). Thompson et al. (1991b) reported that serum carboxylesterases were generally more sensitive to OP exposure then brain acetylcholinesterase. A number of different species have been examined for carboxylesterase activity in response to agrochemical exposure. Of the many studies performed, only a few are discussed here to illustrate the prospects for application to ecosystem-wide monitoring.

2.3.1 Applications in Fish

Work by Wogram et al. (2001) compared the sensitivity of acetylcholinesterase, butyrylcholinesterase, and carboxylesterase in the three-spined stickleback (*Gasterosteus aculeatus*) following exposure to parathion. Results were mixed, with exposure to 0.01 and 0.1 µg/L not affecting activity of any of the three enzymes, although exposures at 1.0 µg/L decreased butyrylcholinesterase activity in the liver (~60%), axial muscle (30%), and gills (30%). No effects were observed on acetylcholinesterase or carboxylesterase activity. Carboxylesterase activity has been reported to be an appropriate biomarker in the fathead minnow (*Pimephales promelas*) (Denton et al. 2003), the spotted gar (*Lepisosteus oculatus*) (Huang et al. 1997), gilthead seabream (*Sparus aurata*) larvae, and the Nile tilapia (*Oreochromis niloticus*) (Pathiratne and George 1998). Ferrari et al. (2007) demonstrated that carboxylesterase inhibition is a good biomarker of exposure to azinphos-methyl and carbaryl in juvenile rainbow trout (*Oncorhynchus mykiss*), with significant carboxylesterase inhibition observed after 24 hr exposure to 2.5 µg/L and 1 mg/L azinphos-methyl and carbaryl, respectively. Work with channel catfish (*Ictalurus punctatus*) indicated that carboxylesterases were inherently more sensitive to inhibition by OP compounds, including chlorpyrifos-oxon, paraoxon, and DEF (*S,S,S*-tributylphosphorotrithioate), but those data did not confirm that carboxylesterases protected against acetylcholinesterase inhibition (Straus and Chambers 1995). However, data from two populations of mosquitofish (*Gamusia affinis*) showed that carboxylesterases exhibited a much higher affinity for OPs than acetylcholinesterases, suggesting a protective role (Chambers 1976). Küster (2005) examined the use of zebrafish (*Danio rerio*) embryos in assessing insecticide exposure using methyl-paraoxon. Data showed that carboxylesterase activity was ~40 times higher then acetylcholinesterase activity at a very early stage of development (6-somite stage), suggesting that carboxylesterases could serve as a stoichiometric buffer system. However, at later stages of development (48 hr postfertilization), the enzyme activities were approximately equal and data showed similar inhibition of the two enzymes following methyl-paraoxon exposure. Therefore, results were inconclusive as to whether carboxylesterase would be superior to acetylcholinesterase for environmental monitoring in this system. Follow-up studies by Küster and Altenburger (2006) concluded that carboxylesterase activity was not preferentially inhibited relative to acetylcholinesterase following methyl-paraoxon exposure. However, the authors made the point that their choice of substrate for carboxylesterase activity assays (*S*-phenylthioacetate) may have confounded the results as this substrate can also be hydrolyzed by paraoxonases (PONs) and arylesteraes (A-esterases) (Küster and Altenburger 2006). *S*-Phenylthioacetate has been used for measuring both carboxylesterase and acetylcholinesterase activity by a number of authors (Bonacci et al. 2004; Corsi et al. 2004; Arufe et al. 2007), which underscores the importance of substrate selection in ensuring an accurate measurement of the desired endpoint.

It has also been reported that both the total carboxylesterase specific activity as well as individual response to OP inhibition are organism specific. A comparison of topmouth gudgeon (*Pseudorasbora parva*), goldfish (*Carassius auratus*), Nile tilapia (*Tilapia nilotica*), mosquitofish (*Cambsua affinis*), and rainbow trout (*Oncorhynchus mykiss*) found that overall activity in the liver was species dependent (Li and Fan 1996, 1997). In addition, the tissue distribution of carboxylesterase activity was species dependent. However, Baron et al. (1999) reported that the carboxylesterase activity of whole-fish homogenates from juvenile rainbow trout

(*Oncorhynchus mykiss*), channel catfish (*Ictalurus punctatus*), fathead minnows (*Pimephales promelas*), and bluegill (*Lepomis macrochirus*) did not vary significantly. Tissue-specific variations were observed, with greater activity in the liver than gills. Results from Wheelock et al. (2005a) supported these findings with assays from three teleost species, Chinook salmon (*Oncorhynchus tshawytscha*), medaka (*Oryzias latipes*), and Sacramento splittail (*Pogonichthys macrolepidotus*), showing that whole-body homogenates demonstrated essentially equivalent carboxylesterase activity. Wheelock et al. (2005a) also examined the dose–response relationship in Chinook salmon following exposure to an OP and pyrethroid insecticide. Exposure to chlorpyrifos at a high dose (7.3 μg/L), but not a low dose (1.2 μg/L), significantly inhibited acetylcholinesterase activity in both brain and muscle tissue (85% and 92% inhibition, respectively), whereas esfenvalerate exposure had no effect in this species. In contrast, liver carboxylesterase activity was significantly inhibited at both the low and high chlorpyrifos doses (56% and 79% inhibition, respectively), while esfenvalerate exposure still had little effect. The inhibition of carboxylesterase activity at levels of chlorpyrifos that did not affect acetylcholinesterase activity suggests that some salmon carboxylesterase isozymes may be more sensitive than acetylcholinesterase to inhibition by OPs.

Carboxylesterase activity has also been used as a biomarker of exposure to non-acetylcholinesterase-inhibiting compounds. For example, the white-spotted rabbitfish (*Siganus canaliculatus*) has been suggested as a biomarker of organotin exposure (Al-Ghais et al. 2000). Taken together, these studies provide a solid base demonstrating that carboxylesterase may be useful for development as a biomarker of exposure to OP and/or carbamate insecticides, as well as other contaminants. However, the inconsistent nature of study results makes direct comparisons difficult, and a great deal of additional data is required before strong conclusions can be drawn.

2.3.2 Applications in Bivalves

Bivalve mussels have been used extensively for environmental monitoring (O'Connor 2002; Sarkar et al. 2006). It has been demonstrated in these organisms that carboxylesterase activity has greater sensitivity then does cholinesterase to inhibition by OPs or carbamates (Ozretic and Krajnovic-Ozretic 1992; Escartin and Porte 1997; Basack et al. 1998; Galloway et al. 2002). Specifically, studies with the mussel *Mytilus galloprovincialis* examined the sensitivity of acetylcholinesterase, butyrylcholinesterase, and carboxylesterase to fenitrothion, fenitrooxon, and carbofuran exposure (Escartin and Porte 1997). Results showed that carboxylesterase activity in the digestive glands and gills was significantly higher than was acetylcholinesterase (~40 fold and 3 fold, respectively) or butyrylcholinesterase activity (~100 fold and 70 fold, respectively). In addition, carboxylesterase activity was much more sensitive to both fenitrothion and fenitrooxon exposure than was acetylcholinesterase activity, although both enzymes were equally sensitive to carbofuran exposure. Similar work in the mussel *Mytilus edulis* indicated that carboxylesterase activity was slightly more sensitive than was acetylcholinesterase to paraoxon and chlorpyrifos exposure (Galloway et al. 2002).

2.3.3 Applications in Crustaceans

Crayfish (*Procambarus clarkia*) are reported to be a useful model species for biomarker studies (Vioque-Fernandez et al. 2007b), but results in field studies did not exhibit direct correlations between pesticide concentrations and esterase activity, suggesting that other factors could affect esterase activity (Vioque-Fernandez et al. 2007a). Field studies were performed with the freshwater crustacean *Asellus aquaticus* (L.) (an isopod) in sites above and below a sewage treatment facility above the Mersey estuary in England (O'Neill et al. 2004). Results showed that both acetylcholinesterase and carboxylesterase activity declined at sites downstream of the sewage effluent discharge; however, carboxylesterase activity was more severely inhibited (27% of control values) and continued to decline at greater distances from the treatment facility.

Studies with the cladoceran *Daphnia magna* have evaluated the use of carboxylesterase activity as a marker of OP exposure. Results showed that carboxylesterase activity was more sensitive to malathion and chlorpyrifos exposure than to acetylcholinesterase, whereas the two enzymes demonstrated equivalent sensitivity to the carbamate carbofuran (Barata et al. 2004). Follow-up field studies in the Delta del Ebro in northeast Spain demonstrated that both acetylcholinesterase and carboxylesterase activity strongly correlated with fenitrothion levels (Barata et al. 2007). Laboratory studies demonstrated that carboxylesterase activity in *Daphnia magna* is sensitive to malathion exposure, with a complete loss in activity following a 15- or 30-min exposure to 0.2 ppm malathion (Bond and Bradley 1997).

2.3.4 Applications in Algae

A number of studies have evaluated esterase activity in algae for use in environmental monitoring. However, the mechanisms by which esterase activity is reduced in algae are unclear. All algae studies cited examined responses in esterase activity following exposure to non-OP or carbamate insecticides. Early work by Blaise and Ménard (1998) demonstrated that esterase activity could be an appropriate indicator of sediment toxicity. This concept was expanded by Regel et al. (2002), who examined esterase activity in two species of algae (*Microcystis aeruginosa* and *Selenastrum capricornutum*) following exposure to acid mine drainage in a South Australian stream. Exposure to acid mine drainage for 1 hr resulted in 30%–70% reduction in esterase activity, which was maintained for 24 hr. Similar studies were performed with seven benthic marine algae, with the diatom *Entomoneis* cf. *punctulata* being the most suitable for assay development. A whole-sediment and water-only toxicity test was developed for algae based upon inhibition of esterase activity, with results showing sensitivity to copper and sediment particles (copper tailings) (Adams and Stauber 2004). Follow-up studies in which *Entomoneis* cf *punctulata* was exposed to hydrocarbon-contaminated sediments showed this assay to have efficacy (Simpson et al. 2007). These authors recommended the use of algal esterase activity as a whole-sediment TIE method to determine the contribution of hydrocarbon contamination to sediment toxicity.

2.3.5 Applications in Terrestrial Organisms

Studies on carboxylesterase activity following exposure to insecticides are not limited to aquatic organisms. Several studies have been performed in terrestrial organisms including birds, lizards, and earthworms. The use of earthworms in environmental monitoring was proposed because earthworms ingest large amounts of soil and are continuously exposed to contaminants through their alimentary surfaces. These applications have recently been reviewed and may be of interest to readers (Sanchez-Hernandez 2006; Castellanos and Sanchez-Hernandez, 2007).

Extensive studies have been performed in avian systems (Thompson 1993). A study comparing plasma cholinesterase and carboxylesterase activity in pigeons (*Columba livia*), American kestrels (*Falco sparverius*), red-tailed hawks (*Buteo jamaicensis*), and Swainson's hawks (*Buteo lineatus*) reported generally lower inhibition of carboxylesterase activity than cholinesterase activity following OP exposure (Bartkowiak and Wilson 1995). Studies aimed at developing a nondestructive biomarker of OP exposure examined blood esterase levels in Japanese quail (*Coturnix coturnix japonica*) and swallows (*Hirundo rustica*) following azamethiphos exposure (Fossi et al. 1992, 1994; Lari et al. 1994). Data showed that carboxylesterase activity was consistently less inhibited then acetylcholinesterase activity, potentially indicating that carboxylesterase activity was a more sensitive endpoint. Carboxylesterase activity in the nestling European starling (*Sturnus vulgaris*) was found to be more sensitive to diazinon exposure then to acetylcholinesterase, but methyl-paraoxon and aldicarb exhibited higher affinities for plasma acetylcholinesterase (Parker and Goldstein 2000). Additional studies in starlings observed a dose–response relationship in serum cholinesterase and carboxylesterase activities following exposure to demeton-*S*-methyl and triazophos (Thompson et al. 1991b). Field studies were conducted with nestling and adult great tits (*Parus major*) following spray drift exposure to pirimicarb and dimethoate (Cordi et al. 1997). Adults demonstrated significant reductions in butyrylcholinesterase activity 24 hr after dimethoate and pirimicarb exposure (51% and 67% of preexposure values, respectively). However, no significant inhibition was observed for serum carboxylesterase activity. Studies with nestlings showed significant decreases in both butyrylcholinesterase and carboxylesterase activity following exposure to dimethoate (66% and 77% of preexposure values, respectively). Interestingly, significant inhibition (27%) was observed in dead nestlings from hedges treated with pirimicarb, whereas no significant reductions in butyrylcholinesterase activity were observed. Interspecies differences in esterases were evaluated in seven species of wild birds to investigate their different susceptibilities to OPs, with results evidencing an inverse correlation between brain acetylcholinesterase and plasma carboxylesterase activity in relationship to body size (Fossi et al. 1996). Carboxylesterase levels were examined in a series of European raptors, with data suggesting that diet affected enzyme activity (Roy et al. 2005). It has been speculated that the wide range of esterases present in the pheasant (*Phasianus colchicus*) may protect it against anticholinergenic pesticides and thereby contribute to its success in regions of the United States where other avian species are adversely affected by pesticides (Baker et al. 1966).

Field studies with the lizard *Gallotia galloti* found a 50% reduction in serum carboxylesterase activity following spray application with parathion (Sanchez et al. 1997a). In addition, laboratory studies with *Gallotia galloti* found that carboxylesterase activity was inhibited for a longer duration than was brain acetylcholinesterase activity following parathion exposure, suggesting that carboxylesterase activity was a superior biomarker (Sanchez et al. 1997b). However, field studies with *Gallotia galloti* in the Canary Islands found that carboxylesterase and acetylcholinesterase activity were essentially equally inhibited following exposure to the OP trichlorphon, whereas butyrylcholinesterase activity was only slightly affected (Fossi et al. 1995). The toad *Chaunus schneideri* was assessed for its ability to serve as an indicator organism for agrochemical exposure (Attademo et al. 2007). Toads were collected in rice fields and surrounding environments and in a reference (control) pristine forest. Carboxylesterase activity in plasma of toads collected from agricultural areas was depressed relative to control values, suggesting that plasma carboxylesterase activity could serve as a nondestructive test for exposure to pesticides.

2.3.6 Miscellaneous Applications

Hamers et al. (2000) reported the development of a small-volume bioassay for quantifying the inhibition potency of OPs and carbamates in rainwater. They compared the use of purified acetylcholinesterase from the electric eel (*Electrophorus electricus*) and carboxylesterase activity from a homogenate of honeybee heads (*Apix mellifera*). Results showed that carboxylesterase activity demonstrated greater sensitivity than acetylcholinesterase activity, with an assay detection limit of "esterase inhibiting potency in rainwater" of 2 ng dichlorvos equivalents per liter. Based upon this assay, they were able to quantify esterase inhibition potential in four rainwater samples, measuring dichlorvos equivalents of 12–125 ng/L. Follow-up studies employed the assay to determine esterase inhibition potency of rainwater collected over 26 consecutive periods, and reported that dichlorvos equivalents exceeded permissible levels in The Netherlands (Hamers et al. 2003). Studies conducted with the lugworm (*Arenicola marina*) found that carboxylesterase levels were significantly lower than those of cholinesterase suggesting to the authors that cholinesterase would be a potentially useful biomarker for this species (Hannam et al. 2007). Studies in the marine worm *Nereis (Hediste) diversicolor* following exposure to temephos showed significant acetylcholinesterase and carboxylesterase inhibition, with carboxylesterase displaying the greatest sensitivity (Fourcy et al. 2002). Carboxylesterase activity has been used in a multiple biomarker study designed to measure the effects of dimethoate exposure in spiders (Babczynska et al. 2006).

Carboxylesterase activity has also been investigated for applications in the selective bioactivation of herbicides in weeds (Gershater et al. 2006). Proteins from a range of important commodity crops and economically important weeds were assayed for carboxylesterase activity. The crops included maize, rice, sorghum, soybean, flax, and lucerne, and the weeds were *Abutilon theophrasti*, *Echinochloa crusgalli*, *Phalaris canariensis*, *Setaria faberii*, *Setaria virdis*, *Sorghum halepense*, and the model plant *Arabidopsis thaliana*. Hydrolysis activity was measured using a range of herbicidal esters including 2,4-D methyl ester, clodinafop propargyl, fenthioprop ethyl, fenoxaprop ethyl, bromoxynil octanoate, and cloquintocet mexyl as well as the insecticide permethrin. Significant hydrolysis of the majority of herbicides was observed, with very few exceptions. Consequently, the applications of carboxylesterase activity in environmental monitoring can potentially be extended to plants and their ability to metabolize ester-containing pesticides. Ileperuma and coworkers (2007) crystallized a carboxylesterase from a kiwifruit species (*Actinidia eriantha*) and showed that it was significantly inhibited by paraoxon, demonstrating that a plant carboxylesterase had a similar inhibitor-binding mechanism as mammalian orthologues. An additional area not covered in this review involves microbial esterases (Bornscheuer 2002), which have important roles in the degradation of agrochemicals (Karpouzas and Singh 2006; Singh and Walker 2006). It is very possible that microbial enzymes could be employed in biomonitoring and bioremediation approaches (Sutherland et al. 2002, 2004), with the particular benefit that it is possible to engineer microbial enzymes with the desired activity and substrate specificity (Yang et al. 2003). Further work should examine the potential to exploit this system.

2.3.7 Future Applications

Taken together, these studies demonstrate a concerted effort to evaluate use of carboxylesterase activity as an indicator of agrochemical exposure. Although results are mixed, existing data suggest that the concept has merit and justifies continued development of this endpoint for use in environmental monitoring studies. However, the use of carboxylesterase activity as a biomarker of exposure will be challenging. Little work has been done to characterize the constitutive levels of esterases in most species, and it is difficult to correlate measured levels of enzyme activity with observed effects. Reduced enzyme activity may either be an indication of enzyme inhibition following exposure to OPs or carbamates or an indication of low esterase expression resulting from other environmental factors. Moreover, very little is known about inducers of carboxylesterase levels. It is possible that some environmental contaminants cause increases in constitutive carboxylesterase levels. It is therefore necessary to develop a method to quantify absolute levels of carboxylesterase. One potential approach is the development of species-specific antibodies, allowing for quantification of the level of carboxylesterase present and subsequent determination of inhibition levels. Because enzyme purification and antibody generation are time intensive, this approach will be expensive. Another potential method is use of reactivation protocols as a biomarker of pesticide exposure to differentiate dilution-reversible inhibitions (carbamate exposure) from dilution-irreversible effects (OP exposure) (Sanchez-Hernandez 2006; Vioque-Fernandez et al. 2007b).

It is also possible that esterase activity could be used as a biomarker of susceptibility, similar to work done with activation of OPs by cytochrome P450s (Keizer et al. 1995). By examining a large numbers of individuals, it could be possible to establish a range of constitutive enzyme levels for endogenous carboxylesterase titers. These levels could then be scored on their ability to hydrolyze pyrethroids, or bind OPs and/or carbamates, with lower scores rated as more sensitive to pyrethroid and/or OP/carbamate toxicity. This process would essentially create an index of susceptibility based upon carboxylesterase activity. One limitation to this concept is illustrated by the fact that rainbow trout sodium channels are more sensitive to pyrethroid toxicity than mouse sodium channels (Glickman and Lech 1982). Therefore, caution must be exercised in interpreting susceptibility based upon constitutive carboxylesterase activity. However, a biomarker of susceptibility to agrochemicals would be useful for examining ecosystem effects and could eventually be employed in risk assessments. Further research should focus on characterizing esterase hydrolytic profiles in a range of species with a goal to examine the relationship to agrochemical toxicity.

An important point in the development of standardized carboxylesterase monitoring protocols will be the selection of the substrate employed for activity measurements (Wheelock et al. 2005a,c). The choice of substrate can profoundly influence observed enzyme activity. The most common substrates currently used include acetyl esters of *p*-nitrophenol (*p*-nitrophenyl acetate, PNPA) or naphthol (α- and β-naphthyl acetate). These substrates are employed because assay endpoints can be measured easily and inexpensively with spectrophotometers. However, the biological significance of these compounds is not known. Some authors have attempted to employ more environmentally realistic substrates, such as pyrethroid surrogate substrates (Riddles et al. 1983; Butte and Kemper 1999; Wheelock et al. 2003; Stok et al. 2004a; Huang et al. 2005, 2006; Devonshire et al. 2007). Although more appropriate measurements of pyrethroid hydrolysis activity, these substrates are not necessarily appropriate for estimating the extent of hydrolysis (i.e., detoxification) of other compounds. In addition, the rates of hydrolysis of pyrethroid surrogate substrates are often so low that they are impractical for many organisms (Wheelock et al. 2005a). It is therefore not appropriate to use a single substrate to characterize esterase activity of crude tissue homogenate. Because multiple esterase isoforms are usually present in the preparation, it is optimal to have a battery of substrates for full characterization of enzyme activity. For example, analyses performed on PNPA and pyrethroid hydrolysis activity in human liver microsomes showed very little correlation between the hydrolytic profiles ($r^2 = 0.29$ for a fenvalerate surrogate), suggesting that different enzymes are involved in the hydrolysis of the two substrates (Wheelock et al. 2003). Therefore, monitoring of PNPA activity, or that of other general substrates, may not provide an accurate account of pyrethroid hydrolysis. Similarly, Stok et al. (2004a) reported that the portion of pyrethroid hydrolysis activity in mouse liver microsomes was only 0.5% that of total esterase activity (as measured by PNPA). Ultimately, it is best to use a suite of substrates for environmental monitoring purposes. The employment of standard substrates with facile esters would enable a theoretical maximum measure of hydrolysis activity as well as comparisons with literature values. However, if such data were combined with more-specific substrates, then a broader measure of the effects of exposure to insecticides and/or other esterase-inhibiting compounds could be performed.

If one is looking at esterase activity in a high-throughput system to evaluate its ability to degrade environmental chemicals, having surrogate substrates is very important because they can lead to inexpensive and quantitative assays that can be routinely performed. For example, an alternative made increasingly attractive by the use of 96- and 384-well plate readers and sophisticated robotics is the use of an array of esterases with varying sensitivity, with substrates such as OPs and carbamates, as an environmental screen for inhibitors of these enzymes (Wortberg et al. 1996). Robotic systems and computer control allow enzymes with different substrates to be used in a high-throughput manner, enabling the rapid screening of multiple surrogate substrates. There are mathematical approaches that can be used to integrate the output of such studies for the tentative identification of environmental contaminants. Such arrays offer a broader range of sensitivity to xenobiotics than any one enzyme. For example, the pattern of inhibition may suggest the composition of mixtures or the identity of a specific inhibitor. In addition, surrogate substrates can be used to drive the purification of esterases from environmental organisms, to guide selection of recombinant esterases in artificial evolution and protein engineering, and for quality control of esterase batches in TIE procedures discussed later. However, it must be remembered that despite their many advantages surrogates are not the substrates of interest. The hypothesis that the substrate used is an appropriate surrogate for the environmental chemical targeted must be tested repeatedly. Also, with the advances in LC-MS and other analytical technologies, it is increasingly attractive to use the actual environmental contaminant of interest rather than or in addition to surrogate substrates.

It is clear that there is interest and need for an increased battery of biomarkers and that a number of researchers are working avidly on the problem (Thompson 1999; Hyne and Maher 2003; Galloway 2006; Galloway et al. 2006; Sarkar et al. 2006). It is likely that carboxylesterase activity will be a valuable addition to methods for determining organism exposure and susceptibility to environmental contaminants. However, it is important to standardize activity assays, ensure that appropriate substrates are used, and make certain that data are interpreted correctly.

3 Pyrethroids

3.1 Introduction

Agrochemical usage practices are currently shifting, with a general movement away from OPs toward pyrethroid insecticides (Casida and Quistad 1998). These trends have been strengthened by passage of The Food Quality Protection Act

(FQPA) in 1996 (Wagner 1997; Glade 1998). The FQPA mandated that EPA consider the "*available information concerning the cumulative effects of such residues and other substances that have a common mechanism of toxicity ... in establishing, modifying, leaving in effect, or revoking a tolerance for a pesticide chemical residue*" (Mileson et al. 1998). Accordingly, the use of OP insecticides is decreasing in California, with a subsequent increase in pyrethroid usage (Epstein et al. 2000). Pyrethroids now account for more than 18% of the world insecticide market (Pap 2003). The ecological implications of this large-scale shift in pesticide use are unknown. Pyrethroids generally have low mammalian toxicity (Abernathy and Casida 1973; Casida et al. 1983; Casida and Quistad 1995), especially when compared to many OP pesticides. However, there have been several reports regarding the sensitivity of aquatic invertebrates and some fish species to pyrethroids (Bradbury and Coats 1989b; Werner et al. 2002; Denton et al. 2003). There is growing concern that the ecological consequences of increased pyrethroid use on aquatic ecosystems could be far reaching.

3.2 *Chemistry*

3.2.1 Background

The pyrethroid insecticides are synthetic analogs of the naturally occurring pyrethrum flowers (Casida 1973; Elliott 1976; Davies 1985). Pyrethrum flowers are of the genus *Chrysanthemum*, of which there are two species, those with red and those with white flowers. Only those with white flowers, *Chrysanthemum cinerariaefolium* Vis, contain the insecticidal active components (Katsuda 1999). The original home of *Chrysanthemum* is the Dalmatian region of the former Yugoslavia, on the Mediterranean coast of the Adriatic Sea (Katsuda 1999). Fujitani (1909) first separated the insecticidal active mixture from pyrethrum flowers and termed the ester component "pyrethron." The structure of "pyrethron acid" was further probed by Yamamoto (1923, 1925) using elegant natural product chemistry, confirming the presence of the cyclopropane ring. Full structures for pyrethrins-I and -II were proposed by Staudinger and Ruzicka (1924), with LaForge and Barthel (1945) reporting that natural pyrethins consisted of four homologues (pyrethrins-I and -II, and cinerins-I and -II). The first pyrethroid, allethrin, was developed under the pressure of World War II by Schechter et al. (1949), first as a mixture of eight isomers from three chiral centers and finally as *S*-bioallethrin (the most active isomer, 1*R*, 3*R*, 4'*S*). The extract of pyrethrum flowers has long been used as an insect control agent but was superseded by the more effective and simpler chlorinated hydrocarbon and OP insecticides after World War II (Casida 1980). However, organochlorines and OPs have subsequently either been eliminated or curtailed because of a panoply of adverse characteristics. Their disappearance has made way for a new generation of pyrethrum derivatives that have improved photostability and selective toxicity, rendering these compounds appropriate for agricultural application. In the

1960s and 1970s, Michael Elliott and coworkers, as well as researchers at Sumitomo Chemical Co., developed a series of synthetic pyrethroids, several of which are still important insecticides, which were the precursors of current pyrethroids (Elliott et al. 1965, 1973a,b, 1974; Ohno et al. 1976). Pyrethroid structures are best described in terms of their distinct acid and alcohol moieties, with synthesis involving condensing these distinct groups to form a connecting ester moiety (essentially the reverse of the hydrolysis reaction shown in Fig. 1). Synthetic efforts focusing on the distinct acid and alcohol moieties were made in an attempt to improve the photostability, insecticidal properties, selective toxicity, and physical characteristics of this class of compounds as well as reduce their overall cost (Casida et al. 1983). The acid portion of the structure was standardized as chrysanthemic acid (Elliott et al. 1965), and then a range of alcohol moieties were tested for insecticidal activity. The instability to sunlight of the 5-benzyl-3-furylmethyl alcohol moiety (resmethrin) (Elliott et al. 1967) led to the inclusion of 3-phenoxybenzyl alcohol in pyrethroid structures (Elliott et al. 1973b). A number of earlier pyrethroids were developed with the original chrysanthemic acid moiety and varying alcohol moieties, including pyrethrin I, allethrin, and resmethrin (Casida 1980). To increase environmental stability, structural modifications in chrysanthemic acid were made including the substitution of the dichlorovinyl acid analogue with chlorine in place of methyl in the isobutenyl side chain (Elliott et al. 1973a) (permethrin; or with dibromovinyl acid to form deltamethrin) (Elliott et al. 1974). Another significant discovery was that the cyclopropane carboxylate moiety could be replaced with the corresponding α-isopropyl 4-chlorophenylacetate (fenvalerate) (Ohno et al. 1976). A major change in the alcohol moiety was also made involving the use of 3-phenozybenzaldehyde cyanohydrin instead of 3-phenoxybenzyl alcohol, to give an α-cyano group-substituted ester (Elliott 1976). This substitution is the distinguishing feature between type I and type II pyrethroids (i.e., permethrin vs. cypermethrin). This change had the effect of converting the ester linkage from a primary to a secondary ester, thereby increasing the chemical stability of the compounds. Additional changes in the alcohol moiety to affect the selectivity and toxicity of the compounds were achieved by adding fluorine substituents to the 3-phenoxybenzylalcohol moiety (cyfluthrin). Pyrethroids contain varying chiral centers and as such can have a variety of optical isomers, often with varying biological activity (Pap 2003). For example, there are eight different isomers of cypermethrin. The *S,S*-isomer of fenvalerate has substantial efficacy relative to the *R,R*-isomer or the racemic mixture, leading to the selective manufacture and sale of the *S,S*-isomer (esfenvalerate vs. fenvalerate). Accordingly, it is important that chirality be taken into account when analyzing different pyrethroids (Soderlund and Casida 1977).

3.2.2 Hydrophobicity

Pyrethroids exhibit a high degree of hydrophobicity, as shown in Table 1. The logs of the octanol–water partition coefficients (log K_{ow} or log P) are generally on the order of 5 or greater. This physical characteristic has a number of environmental

Table 1 Structure and physical constants of common pyrethroids[a]

Structure	Name[b] (type)[c]	Water solubility[d]	Vapor pressure[e]	Henry's constant[f]	Log P[g]
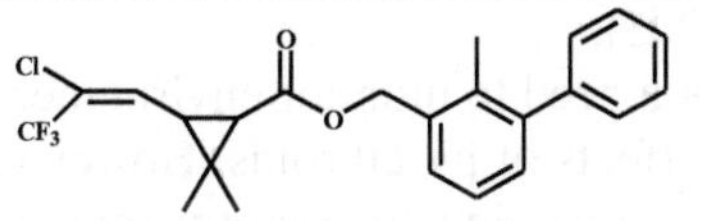	Bifenthrin (I)	0.014	2.4×10^{-2}	7.2×10^{-3}	6.40
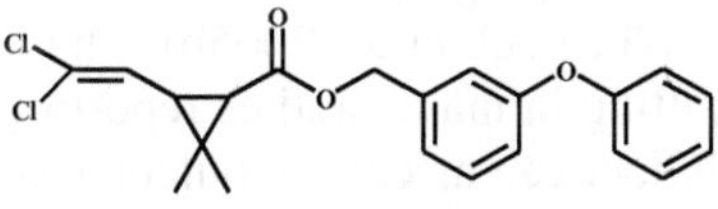	Permethrin (I)	5.5	2.5×10^{-3}	1.4×10^{-6}	6.10
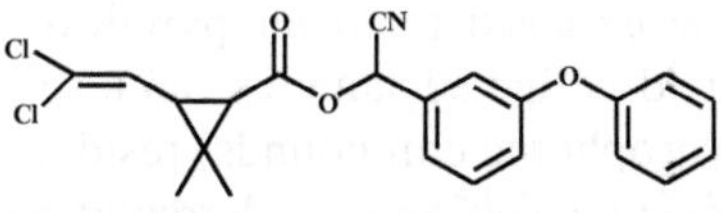	Cypermethrin (II)	4.0	2.0×10^{-4}	3.4×10^{-7}	6.54
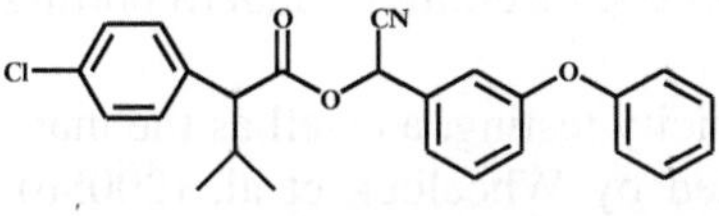	Fenvalerate (II)	6.0	1.92×10^{-2}	1.4×10^{-7}	5.62
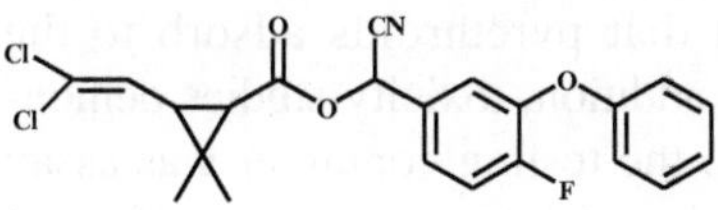	Cyfluthrin (II)	2.3	<1	3.7×10^{-6}	5.97
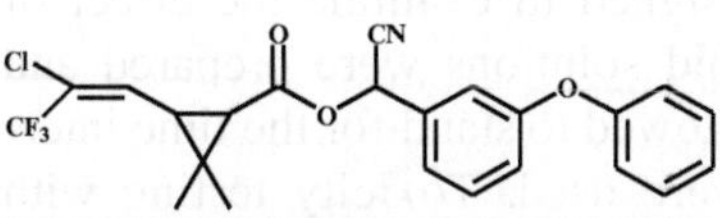	λ-Cyhalothrin (II)	5.0	1.0×10^{-3}	1.9×10^{-7}	7.0

[a]Data are from Laskowski (2002).
[b]Chemical names of each pesticide are as follows: permethrin ((3-phenoxyphenyl) methyl 3-(2,2-dichloroethenyl)-2,2-dimethylcyclopropane carboxylate, CAS 52645-53-1), bifenthrin ((2-methyl[1,1-biphenyl]-3-yl)methyl (1*R*,3*R*)-3-[(1*Z*)-2-chloro-3,3,3-trifluoro-1-propenyl]-2,2-dimethylcyclopropanecarboxylate, CAS 82657-04-3), cypermethrin ((+/−) α-cyano(3-phenoxyphenyl)methyl 3-(2,2-dichloroethenyl)-2,2-dimethylcyclopropanecarboxylate, CAS 52315-07-8), fenvalerate ((*R/S*)-α-cyano(3-phenoxyphenyl)methyl (α*R/S*)-4-chloro-(1-methylethyl) benzeneacetate, CAS 51630-58-1), cyfluthrin (α-cyano(4-fluoro-3-phenoxyphenyl)methyl 3-(2,2-dichloroethenyl)-2,2-dimethylcyclopropanecarboxylate, CAS 68359-37-5), λ-cyhalothrin ((*S/R*)-α-cyano(3-phenoxyphenyl)methyl (1*S*,3*S*)-3-[(1*Z*)-2-chloro-3,3,3-trifluoro-1-propenyl]-2,2-dimethylcyclopropanecarboxylate, CAS 91465-08-6).
[c]Type II pyrethroids contain an α-cyano group on the benzylic carbon (see Fig. 1).
[d]Units are μg/L.
[e]Units are mPa.
[f]Units are atm m^3 mol^{-1}.
[g]Values are the log of the partitioning constant between octanol and water (K_{ow}).

ramifications. It was initially thought that the low aqueous solubility of these compounds would prevent their runoff from sites of agricultural application. This issue can be of particular importance in California during dormant spraying, which often coincides with winter storm events (Werner et al. 2002, 2004; Teh et al. 2005). Migration of pyrethroids from soils following application has been observed in association with particulate material. It appears that pyrethroids adsorb to sediments, which can then be washed offsite, through agricultural drainage ditches and eventually into larger waterways (Gan et al. 2005). There is still concern for the environmental impact of pyrethroids that leach from sites of application into aquatic ecosystems (Gan et al. 2005).

The growth in pyrethroid usage has resulted in a need to monitor environmental samples for the presence and potential toxic effects of pyrethroids. However, the hydrophobicity of pyrethroids can make it challenging to perform toxicity assays with aquatic samples (Lee S. et al. 2002; Wheelock et al. 2005b). These compounds readily adsorb to test containers, resulting in many studies reporting aqueous levels as "nominal concentrations." This fact can affect the outcome of toxicity testing results in that organisms may be exposed to lower pyrethroid concentrations than those targeted. In addition, field-collected samples can lose a large proportion of their pyrethroid (and other hydrophobic compounds) residues through adsorption to sampling and testing containers, resulting in underreporting of observed pyrethroid toxicity.

The effects of such pyrethroid loss during toxicity testing, as well as the magnitude of loss to sampling containers, was tested by Wheelock et al. (2005b). Results with different container types showed that pyrethroids adsorb to the container surface in a time-dependent manner. In addition, toxicity studies demonstrated that the effect of pyrethroid adsorption to the testing container was assay condition dependent. Toxicity studies were designed to examine the effect of sample incubation on toxicity. Spiked pyrethroid solutions were prepared and then added to the sample testing container and allowed to stand for the time intervals indicated in Table 2. Two test species were used. Toxicity testing with *Ceriodaphnia dubia* showed very distinct time-dependent adsorption effects upon observed toxicity with an approximate 50% reduction in toxicity between the 30-min and 4-hr incubation, whereas *Hyalella azteca* did not exhibit a significant change. Of particular importance is the observation that vortexing of the *C. dubia* samples after a 4-hr incubation resulted in almost complete recovery of pyrethroid-associated toxicity. These data suggest that the time-dependent loss of toxicity is in fact an adsorption phenomenon.

To further examine the time-dependent effects upon pyrethroid toxicity, similar studies were performed with fathead minnows (*Pimephales promelas*). These studies showed a slight effect of adsorption upon the observed toxicity between the 30-min incubation and subsequent incubations. An important observation from these studies is the effect of sample stirring upon pyrethroid toxicity. The data show that a 24-hr incubation of the pyrethroid-spiked sample before organism addition resulted in only 30% mortality relative to the 30-min incubation. However, simple stirring of the sample with a glass rod for 2 min was sufficient to restore the observed mortality to 100%. These data are important in that sample handling can potentially have profound effects upon the outcome of toxicity assays. Unfortunately, no studies to date have identified a testing method that can prevent systemic pyrethroid loss during sample collection and handling. It is therefore necessary, at a minimum, that methods rigorously describe how samples were manipulated. Sample shaking and stirring

Table 2 Effect of time to test initiation and sample handling on acute permethrin toxicity

Time (hr)[a]	Permethrin concentration (ng/L)[b]				LC_{50}[c]	95% CI[d]
Hyalella azteca[e]						
	0	25	50	75	LC_{50}	95% CI
0.25	13 ± 12	13 ± 12	87 ± 12	87 ± 12	40	34–46
0.5	6 ± 10	27 ± 23	73 ± 23	87 ± 12	35	28–45
1	7 ± 12	13 ± 12	93 ± 12	100 ± 0	35	33–38
2	13 ± 12	33 ± 23	73 ± 12	93 ± 12	38	29–48
4[f]	13 ± 12	0 ± 0	53 ± 12	93 ± 12	48	41–55
Ceriodaphnia dubia[g]						
	0	125	250	375	LC_{50}	95% CI
0.25	0±0	90 ± 12	100 ± 0	100 ± 0	66	61–78
0.5	5±10	85 ± 19	100 ± 0	100 ± 0	74	55–106
1	0±0	80 ± 16	100 ± 0	100 ± 0	78	58–107
2	0±0	70 ± 26	100 ± 0	100 ± 0	89	58–146
4[h]	0±0	40 ± 23	100 ± 0	100 ± 0	140	106–168
Pimephales promelas[i]						
	0	2000	4000	8000	LC_{50}	95% CI
0.5	0	0	5	100	5900	5400–6300
1	0	0	0	60	7300	—[k]
2	0	0	0	100	6000	6000–6000
4	0	0	0	75	6700	—
4/stirring[j]	0	0	0	100	6200	5300–7300
8	0	0	0	90	6200	5300–7300
8/stirring	0	0	0	90	6200	5300–7300
24	0	0	0	30	>8000	—
24/stirring	0	0	0	100	6000	6000–6000

[a]Test solutions were prepared at given permethrin concentrations and placed into test containers followed by organism addition at the time intervals indicated.
[b]Nominal water concentration.
[c]LC_{50} (concentration to cause 50% lethality) values are reported in ng/L. Values were calculated using Spearman–Karber analysis.
[d]95% confidence interval (CI) of the LC_{50} results.
[e]Three replicates of five *Hyalella azteca* used per treatment. Data are given as 96-hr % mortality. Results are from Wheelock et al. (2005b).
[f]Vortexing of the 4-hr sample did not affect the observed toxicity (data not shown).
[g]Four replicates of five neonate *Ceriodaphnia dubia* per treatment. Data are given as 48-hr % mortality. Results are from Wheelock et al. (2005b).
[h]Vortexing of the 4-hr sample served to increase the toxicity to nearly initial levels with 48-hr mortality being 85% ± 15% (data not shown).
[i]Fathead minnows obtained from Aquatox, Inc. (Hot Springs, AK) were maintained in EPA moderately hard (EPAMH) water until tested when 7 d old. Each test sample was tested using two replicates of 10 fish each in 400-mL glass beakers containing 250 mL test solution. Permethrin exposures were conducted at 2000, 4000, and 8000 ng/L in EPAMH water. Permethrin was obtained from Accustandard (New Haven, CT), and working standards (100 µg/L) were prepared in high-performance liquid chromatography-grade methanol. Methanol concentration in all test solutions was less than 0.1%. Fish were added to the test solutions 0.5, 1, 2, 4, 8 or 24 hr after permethrin addition. An aliquot of the 4-, 8-, and 24-hr solutions was vigorously stirred for 2 min before fish addition. Test duration was 96 hr, and test solutions were renewed at 48 hr. Fish were fed *Artemia nauplii* 4 hr before sample renewal at 48 hr. Tests were conducted at 25° ± 1°C with a 16 hr light:8 hr dark photoperiod. Mortality was noted daily. The concentration required to cause 50% mortality

(continued)

Table 2 (continued)

(LC_{50}) in each of the treatments was calculated from the mortality data using a computer program (ToxCalc, Ver. 5.0.23; Tidepool Scientific, McKinnleyville, CA). Standard deviations were not calculated because data were only collected in duplicates.
[j]Test containers were stirred vigorously for 2 min before organism addition.
[k]95% CI values could not be determined from the mortality data.

is of special importance. As demonstrated with the fathead minnow studies, stirring of the sample is sufficient to recover some of the pyrethroid-associated toxicity. It is assumed that the turbulence created during the stirring process is sufficient to create enough force to overcome the hydrophobic associations between pyrethroids and containers. The difficulty will be in developing a reproducible method for resuspending pyrethroids during sample handling. It is therefore extremely important that these effects be considered when performing toxicity assays. It is likely that a number of different toxicity testing systems are affected by the observations presented in Table 2, which could have important ramifications in environmental toxicity testing.

3.3 *Toxicology*

Pyrethroid toxicity varies greatly with species and pyrethroid type (Bradbury and Coats 1989b). The LD_{50} for pyrethroids are structure-, organism-, and life stage dependent. For example, the oral LD_{50} for deltamethrin in adult rats is 81 mg/kg versus 5.1 mg/kg in weanling rats (Sheets et al. 1994). Pyrethroids are primarily sodium channel toxins that prolong neural excitation, but they exhibit few or no direct cytotoxic effects (Casida et al. 1983). The major site of action of all pyrethroids is the voltage-dependent sodium channel (Narahashi 1996); however, a number of other potential interaction sites exist, such as the voltage-gated chloride channels (Forshaw et al. 1993), GABA-gated chloride channels (Bloomquist et al. 1986), and possibly protein phosphorylation (Enan and Matsumura 1993). The degree of sodium channel excitability is dose related, but the nature of the excitability is structure dependent (Coats 1990). Most pyrethroids exhibit much greater toxicity to insects than to mammals because insects have increased sodium channel sensitivity, lower body temperature, a lipophilic cuticle, and smaller body size (Bradberry et al. 2005). In addition, mammals are protected by poor dermal absorption of pyrethroids and rapid metabolism to nontoxic metabolites (Bradberry et al. 2005).

3.3.1 Human Toxicity

Pyrethroids generally exhibit low mammalian toxicity (Vijverberg and van den Bercken 1990; Ray and Forshaw 2000). Despite their extensive worldwide use, there are relatively few reports of human pyrethroid poisoning (Ray and Forshaw

2000). Fewer than 10 deaths have been reported from ingestion or following occupational exposure (Bradberry et al. 2005). In adults, 10 mg/kg oral doses have been reported to cause seizures (Tippe 1993), and some pyrethroids have been reported to cause systemic occupational poisoning in China (He et al. 1989). However, pyrethroids are rapidly hydrolyzed in the liver, thereby preventing the nervous system effects that are lethal to insects (Aldridge 1990). Many pyrethroids are less toxic to mammals than are the very safe natural pyrethrins. However. some of the more stable pyrethroids can give symptoms following occupational exposure. Pyrethrum formulations extracted from chrysanthemum flowers can also contain sesquiterpene lactones, which can cause allergic rhinitis and contact dermatitis (O'Malley 1997). However, processes to remove these components were developed many years ago, and synthetic pyrethroids do not contain these natural by-products.

The main routes of pyrethroid metabolism are through the action of esterases (see Fig. 1) and cytochrome P450s (see Fig. 5). As shown in Fig. 5, P450s can hydroxylate pyrethroids in a number of positions, thereby increasing their water solubility and providing chemical moieties for further conjugation and Phase II metabolic processes (Casida and Quistad 1995). Esterases hydrolyze pyrethroids to the corresponding alcohol and acid, and esterase activity correlates with pyrethroid-associated toxicity (Abernathy and Casida 1973). Further detail on metabolism of pyrethroids is beyond the scope of this review, but readers interested in this topic are referred to a number of available reviews (Bradbury and Coats 1989b; Aldridge 1990; Coats 1990; O'Malley 1997; Ray and Forshaw 2000; Bradberry et al. 2005).

3.3.2 Environmental Toxicity

Pyrethroids exhibit much greater toxicity to aquatic organisms than to mammals (Coats et al. 1989). Laboratory tests have shown that pyrethroids are extremely toxic to fish such as fathead minnow, rainbow trout, brook trout, bluegill, and sheepshead minnow (TDC 2003; Bradbury and Coats 1989a). Pyrethroids have negative temperature coefficients of toxicity, which means that their toxicity increases in colder water (Ware and Whitacre 2004). Pyrethroids are acutely toxic to aquatic insects and crustaceans, with most median lethal concentrations (LC_{50}s) well below 1 µg/L (TDC 2003). Anderson et al. (2006a) reported freshwater bifenthrin and permethrin LC_{50}s below 1 µg/L for the amphipod *H. azteca* (9.3 and 21.1 ng/L) and the mayfly genus *Procloeon* (84 and 90 ng/L); however, the LC_{50}s for the dipteran *Chironomus dilutus* were 6,000 and 10,000 ng/L, respectively. Wheelock et al. (2004) reported *C. dubia* LC_{50}s for five pyrethroids that ranged from 140 to 680 ng/L. Because pyrethroids quickly adsorb to laboratory exposure chambers, the reported thresholds probably underestimate the actual sensitivity of water column organisms (Wheelock et al. 2005b). It should also be stressed that the majority of laboratory-based toxicity studies are aqueous-only exposures that do not take into account the effects of organic material on observed toxicity. It is distinctly possible that the presence of organic material in the testing system will significantly reduce the observed acute toxicity (Bondarenko et al. 2006; Yang et al. 2006a,b,c).

As previously stated, pyrethroids are hydrophobic chemicals that have a tendency to associate with sediment or humic particles. It is therefore more likely to detect pyrethroids in suspended and bedded sediments, and there is greater potential for environmental impacts in this compartment. Maund et al. (2002) reported that the toxicity of cypermethrin-containing sediments was dependent on sediment organic carbon content, with 10-d LC_{50}s as low as 3.6 mg/kg for the freshwater amphipod *H. azteca* and 13 mg/kg for the freshwater dipteran *Chironomus tentans*. Amweg et al. (2005) reported *H. azteca* LC_{50}s for several pyrethroids based on toxicity tests with spiked sediments containing various amounts of organic carbon that ranged from 6.6 ng/g for bifenthrin to 249 ng/g for permethrin. Anderson et al. (2007a) reported several pyrethroid LC_{50}s for two marine amphipod species. The LC_{50}s for bifenthrin, cypermethrin, and permethrin were 0.008, 0.011, and 0.140 mg/kg, respectively, for *Eohaustorius estuarius*, and 0.95, 0.47, and 8.9 mg/kg, respectively, for *Ampelisca abdita*.

Several sublethal effects have also been reported. Denton (2001) reported impacts on fish, including behavioral changes such as rapid gill movement, erratic swimming, altered schooling activity, and swimming at the water surface. Concentrations of pyrethroids as low as 10 ng/L reduced daphnid reproduction and lowered feeding filtration rates (Day 1989). Cypermethrin exposures below 0.004 μg/L significantly impaired salmonid olfactory responses, which could disrupt reproductive functions (Moore and Waring 2001). Several pyrethroids and their breakdown products were found to have endocrine activity (Tyler et al. 2000). For a comprehensive review of the effects of pyrethroids in field studies, see Oros and Werner (2005).

Fish demonstrate extreme sensitivity to pyrethroids (Bradbury and Coats 1989a), which is thought to be partly the consequence of their slow metabolism of the parent compound (Denton et al. 2003). The sensitivity of many fish to pyrethroid application has led to the development of so-called "fish-safe" derivatives including cycloprothrin, etofenprox, flufenprox, and silafluofen (Pap 2003). These compounds exhibit significantly decreased toxicity toward rainbow trout; for example, the LC_{50} for silafluofen is > 100,000 μg/L. However, of the few studies that have reported pyrethroid metabolism in piscine species, esterase measurements were generally made only on tissue pools, rather than on individual fish (Glickman and Lech 1981; Glickman et al. 1982). Although useful, these data only provide information on the average enzyme activity in a population or species and do not indicate the activity range among individuals. Data derived from measurements in individuals are important to determine if some individuals metabolize pesticides slower than others, which could potentially correlate with increased sensitivity to pyrethroid or OP exposure. It has been hypothesized that fish with lower levels of esterase activity are more sensitive to pyrethroid and OP toxicity (Wheelock et al. 2005a). Understanding the range of esterase activity in environmentally sensitive species will be useful for interpreting the impact of increased pyrethroid usage upon aquatic ecosystems.

3.4 Presence in Ambient Samples

Because of their hydrophobicity, pyrethroids do not remain in surface water, but quickly partition onto particulate matter in the environment (Liu et al. 2004). Although the bioavailable concentration of pyrethroid pesticides in the water column is reduced by adsorption, toxic concentrations are sometimes present (Spurlock et al. 2005). A number of California Department of Pesticide Regulation (DPR) studies have detected pyrethroids in California surface waters (Bacey et al. 2003, 2005; Kelley and Starner 2004), but there are few studies of pyrethroids in surface waters outside of California (Schulz 2004). Of the studies conducted other than in California, Schulz (2004) reviewed 15 studies that detected a variety of pyrethroids in surface waters. Five of these studies were conducted in the southern United States and 10 were conducted in Europe and South Africa.

In the DPR studies, total pyrethroid concentrations were measured, but Liu et al. (2004) estimated that dissolved concentrations of pyrethroids in stream waters would be less than 1% of the total. Although the dissolved concentrations in stream water were quite low, the authors demonstrated that the dissolved concentrations of pyrethroids in runoff from a nursery operation were as high as 27% of total residues. Spurlock et al. (2005) tested a probabilistic screening model for dissolved pyrethroid concentrations and determined that although pyrethroids in the water column have reduced bioavailability in the presence of suspended sediment, toxic concentrations could still be present.

Toxic concentrations of pyrethroids are more readily found in bedded sediments. Weston et al. (2004, 2005) detected pyrethroids in sediments associated with agricultural drainages and suburban creeks in central California. Amweg et al. (2006) detected pyrethroids in urban creeks in the San Francisco Bay area of California. Using the toxic unit approach, these authors implicated pyrethroid pesticides as the cause of observed sediment toxicity in their studies. Studies conducted by the University of California Davis have also detected pyrethroids in the sediments of agriculturally dominated creeks and urban drainages and have identified pyrethroids as the cause of observed toxicity through TIE methods (Anderson et al. 2006b, in press; Phillips et al. 2006).

Pyrethroid pesticides have not been routinely measured in large-scale environmental monitoring programs. The United States Geological Survey National Water Quality Assessment Program is the largest urban monitoring program and only screens for permethrin (www.usgs.gov). Some regional studies have measured pyrethroids in water and sediment at concentrations that are toxic to resident organisms. The Surface Water Ambient Monitoring Program (SWAMP) now includes routine analyses of pyrethroids in California watersheds and has detected pyrethroids in surface water and sediment (http://www.waterboards.ca.gov/swamp/index.html). Recent advances in LC-MS technology should reduce the development time and cost of analyzing for multiple pyrethroids in ecosystems. Also, immunoassays that are compound and pyrethroid class selective have been developed for multiple pyrethroids and their key metabolites (Wengatz et al. 1998; Shan et al. 1999, 2000; Watanabe et al. 2001; Lee et al. 2002, 2004; Mak et al. 2005). These low-cost assays should facilitate environmental monitoring.

4 Toxicity Identification Evaluations (TIEs)

4.1 Overview

In recent years, there has been increased emphasis on the use of routine aquatic organism toxicity tests to measure ambient water quality (Bailey et al. 1995, 1996, 2000; de Vlaming et al. 2000; Werner et al. 2000). Studies have been conducted to assess receiving water impacts of discharged storm water (Bailey et al. 1997; Werner et al. 2002), municipal and industrial effluent (Bailey et al. 1995), and agricultural runoff (de Vlaming et al. 2000; Werner et al. 2000; Anderson et al. 2002). Studies have also been conducted to determine the environmental impact of contaminated sediments (Anderson et al. 2001; Hunt et al. 2001). Concomitantly, increased attention has focused on methods for identifying the chemical(s) that are responsible for the toxicity so that appropriate control measures can be taken. The United States Environmental Protection Agency (USEPA) has published a series of TIE methods designed to identify the causes of observed toxicity in aqueous samples using chemical characterization, identification and confirmation procedures (USEPA 1991, 1993a,b, 1996). Additionally, sediment TIE procedures will soon be published (Anderson et al., 2007b; in press; 2007 USEPA).

TIE methods have been broadly applied to identify the causes of toxicity in water and sediment. TIE testing has routinely identified OP insecticides, including diazinon and chlorpyrifos, as causes of toxicity in municipal effluents, surface waters, and sediments in Northern California (Bailey et al. 2000; de Vlaming et al. 2000; Werner et al. 2000; Anderson et al. 2003; Hunt et al. 2003; Phillips et al. 2004). As discussed previously, pyrethroids are difficult to identify using standard TIE methods because of their physicochemical properties (Sharom and Solomon 1981; Casida et al. 1983; Wheelock et al. 2005b) and the lack of inexpensive, sensitive, and selective analytical methods capable of detecting these insecticides at biologically relevant concentrations (Leng et al. 1999; Lee et al. 2002). In addition to the standard USEPA TIE treatments, there are several TIE treatments that can provide specific lines of evidence for pyrethroid toxicity: carboxylesterase addition (Wheelock et al. 2004, 2006), addition of piperonyl butoxide (USEPA 1991, 1993b; Amweg and Weston 2007), and temperature reduction (Weston 2006). These procedures satisfy the criteria for widespread applicability in that they are rapid, relatively inexpensive, and do not require a high level of expertise or expensive equipment. Although PBO addition and temperature reduction help to characterize pyrethroid-associated toxicity, reduction of toxicity with the addition of carboxylesterase provides a stronger line of evidence when identifying pyrethroids as the cause of toxicity.

4.2 TIE Procedures

The generic TIE protocols are performed in three phases: toxicity characterization (Phase I), toxicant identification (Phase II), and toxicant confirmation (Phase III). Table 3 provides a basic list of standard and emerging Phase I and II TIE methods for water and sediment. Each of these TIE treatments are applied to the test sample and comparison of the level of baseline toxicity with the TIE treatments identifies the physicochemical characteristics of the toxicants. It is essential that proper controls and blanks be used with each TIE treatment, and that a high level of QA/QC is maintained throughout the TIE process. If the standard suite of Phase I treatments are ineffective in identifying cause(s) of toxicity, other techniques can be used, including anion- and cation-ion exchange resins and activated charcoal molecular sieves (Burgess et al. 1997). Application of the TIE process has demonstrated its applicability to virtually every test species and in a variety of test matrices.

The EPA Phase I TIE procedures (USEPA 1991, 1992, 1996) describe a process in which the sample is split into aliquots, each of which is subjected to a single TIE manipulation concurrently with the other treatments ("parallel" treatment approach). However, EPA points out that the Phase I TIE characterization procedures are relatively broad and can indicate more than one class of toxicity. Additional tests or an altered approach may be needed to delineate/confirm the role of a particular chemical class in the effluent toxicity, especially when multiple toxicants are present (USEPA 1993a,b). For example, when the primary toxicant is present in high concentrations, it may mask the other potential toxicant(s) in the sample; ammonia is a common example. In these cases, sequential treatments ("stacked" treatment approach) can be used to evaluate the role of secondary toxicants, e.g., removal of ammonia by Zeolite followed by removal of nonpolar organics by solid-phase extraction (SPE) treatment in cases where multiple toxicants are present at toxic concentrations.

Results of Phase I can be compared with pretreatment program data and chemical-specific data to identify potential toxicants. However, chemical analyses conducted in the absence of Phase I TIE information, e.g., chemical class of toxicant(s), to guide the type of analysis are usually wasted expenditures. For this reason, EPA cautions that chemical-specific tracking should be conducted after the toxicant(s) is(are) identified and confirmed in Phase II and Phase III TIEs, respectively (USEPA 1993a,b).

Phase I characterization provides information on the chemical classes responsible for toxicity and is applicable to both acute and chronic endpoints. Following is a brief description of the aquatic treatments that can be used for water column and interstitial water samples (see Table 3 for a summary). Ammonia toxicity can be assessed by Zeolite removal or the graduated pH test. Toxicity resulting from the presence of divalent cation metals can be detected via the addition of ethylenediaminetetraacetic acid (EDTA), and sodium thiosulfate (STS) is added to reduce toxicity from oxidants and specific metals. Contaminants associated with particles are removed by filtration or centrifugation (2500 *g* at 4°C). Volatile constituents

Table 3 Brief description of some standard and emerging phase I and II Toxicity Identification Evaluation (TIE) methods for water and sediment

Sediment treatment	Description	Citations
Zeolite	Addition of 10%–20% zeolite to sediment reduces interstitial and overlying water ammonia	Besser et al. 1998; Burgess et al. 2003
Chelating resin (SIR-300)	Addition of 10% SIR-300 chelates heavy metal ions and reduces metal bioavailability	Burgess et al. 2000
Coconut charcoal (PCC)	Addition of 10%–15% PCC to sediment reduces toxicity caused by organic contaminants	Lebo et al. 1999, 2003; Ho et al. 2002, 2004
Ambersorb 563	Addition of 10% Ambersorb to sediment reduces toxicity caused by organic contaminants	Kosian et al. 1999; Lebo et al. 1999; West et al. 2001
Carboxylesterase	Addition of enzyme to overlying water reduces toxicity caused by pyrethroid pesticides by breaking down compounds into nontoxic forms	Wheelock et al. 2004, 2006; Weston and Amweg, 2007
Piperonyl butoxide (PBO)	Addition of PBO to overlying water can reduce toxicity caused by organophosphate pesticides, and increase toxicity caused by pyrethroid pesticides and DDT	USEPA 1991; Kakko et al. 2000
Temperature reduction	Testing warm water organisms at colder temperatures can increase the toxicity of pyrethroids and DDT	Ware and Whitacre 2004
Water treatment	**Description**	**Citations**
Aeration	The sample is aerated for 1 hr to determine if toxicity is caused by volatile compounds or surfactants	USEPA 1991, 1993b, 1996
Filtration or centrifugation	Reduces toxicity that is particle related; also used as a pretreatment step for the column treatments	USEPA 1991, 1996
EDTA[a]	Organic chelating agent that preferentially binds with divalent metals, such as copper, nickel, lead, zinc, cadmium, mercury, and other transition metals to form nontoxic complexes	USEPA 1991, 1996
Sodium thiosulfate	Reduces toxicity caused by oxidants and some cationic metals	USEPA 1991, 1996
pH adjustment and volatilization	Reduces ammonia in sample by converting it to the unionized fraction through pH adjustment and volatilizing it by stirring	USEPA 1991, 1993b; Burgess et al. 2003

Graduated pH	Determines if pH-dependent toxicants are responsible for the observed toxicity; the toxicity of ammonia, sulfide, and some metals changes with pH	USEPA 1991, 1993b, 1996
Zeolite column solid-phase extraction	Reduces toxicity caused by ammonia	USEPA 1991; Burgess et al. 2004
Cation column solid-phase extraction and elution	Removes metals from the sample; column can be eluted with 1 N hydrochloric acid (HCl) and resulting eluate tested to determine if substances removed by the column were toxic	USEPA 1993b, 1996; Burgess et al. 1997
Nonpolar organic column solid-phase extraction and elution	Removes nonpolar organic compounds; column can be eluted with solvent to determine if substances removed by the column were toxic	USEPA 1991, 1993b, 1996
Carboxylesterase	Addition of enzyme to sample reduces toxicity caused by pyrethroid pesticides by breaking down compounds into nontoxic forms	Wheelock et al. 2004, 2006
Piperonyl butoxide (PBO)	Addition of PBO can reduce toxicity caused by organophosphate pesticides and increase toxicity caused by pyrethroid pesticides and DDT	USEPA 1991; Kakko et al. 2000
Temperature reduction	Testing warm water organisms at colder temperatures can increase the toxicity of pyrethroids and DDT	Ware and Whitacre 2004

[a]Ethylene diaminetetrachloracetic acid.

such as sulfide are oxidized or volatilized by sample aeration. Nonpolar organic compounds are removed/detected by passing the water samples over a C8 or C18 SPE column. The SPE column can then be eluted with methanol or other solvents. A portion of the eluting solvent is then added back to the laboratory control water to determine if the SPE-bound organic compounds cause toxicity. A cation-exchange column is used to remove metal contaminants, and the column is then eluted with 1 N HCl to add back bound metals to clean dilution water. A sequential treatment using the SPE columns is conducted to resolve mixtures of organic and metal contaminants. The presence of metabolically activated compounds such as OPs is examined through the addition of PBO, which should result in a decrease in the toxic signature. An increase in toxicity following PBO addition usually indicates the presence of pyrethroids (Casida 1970; Casida and Quistad 1995; Ameg and Weston 2007).

The Phase II guidance manual (USEPA 1993b) describes procedures for use in identification of specific classes of toxicants, including: ammonia, cationic metals, polar and nonpolar organic chemicals, chlorine, and filterable toxicants. Phase II treatment techniques are similar to Phase I and are applicable to most acute and chronic test methods. Phase II incorporates chemical-specific analytical procedures, including gas chromatography (GC), GC/mass spectrophotometry (GC/MS), high-performance liquid chromatography (HPLC)/MS, atomic absorption (AA), or ion-coupled plasma (ICP)/MS to identify toxicants. Interested readers are referred to the EPA Phase II manual (USEPA 1993b), and to Waller et al. (2005) for a detailed description of Phase II TIE procedures and examples of TIE case studies.

Phase III TIE procedures involve a thorough confirmation of the cause(s) of toxicity and constitute a key part of the TIE process. Suspected toxicant(s) identified in Phase I and Phase II are confirmed through application of one or more Phase III steps, including correlation approach, symptom approach, species sensitivity approach, spiking approach, and mass balance approach (USEPA 1991, 1992, 1993a,b, 1996)

Sediment TIEs are also becoming more common and are more appropriate for pyrethroids because of their hydrophobicity (Anderson et al., in press). The following solid-phase TIE treatments may be used in conjunction with the interstitial water treatments described earlier (Anderson et al., in press). Solid-phase treatments include amending the sediment with substances that reduce toxicity caused by particular classes of chemicals or adding substances to the overlying water in the exposure chamber. As with the aquatic methods, all three phases of the TIE process can be utilized for sediments. Phase I sediment amendments include addition of a carbonaceous resin (e.g., Amberlite XAD4) or of powdered coconut charcoal to reduce the bioavailability of organic chemicals. Cation-chelating resin (e.g., SIR-300) is added to reduce the bioavailability of cationic metals. Zeolite is added to remove unionized ammonia. Overlying water treatments include the addition of carboxylesterase enzyme and bovine serum albumin (BSA) in separate treatments to identify toxicity from pyrethroid pesticides, and the addition of PBO to differentiate between pyrethroid and OP pesticides (see following for descriptions of the carboxylesterase and albumin methods). Phase II treatments include the separation and elution of the Amberlite and SIR-300 resins. The eluate is added to control water to verify that chemicals sorbed to the resins were eluted at toxic concentrations. Phase III procedures include comparing concentrations of chemicals in sediments or in the solvent eluates to known toxicity thresholds.

4.3 TIE Use in Ambient Waters and Sediments

TIEs on ambient waters and sediments are used to determine the cause of sample toxicity, for example, as part of the development of a Total Maximum Daily Load evaluation (TMDL). When toxicity tests or other biological indicators produce evidence of water quality impairment caused by contaminated sediments, the water bodies in which the sediments occur may be placed on the Clean Water Act §303[d] list of impaired waters. For each water body on the §303[d] list, states are required to develop individual TMDLs capable of identifying the cause of impairment, locating all sources of the causative pollutant, and allocating loadings of the pollutant among the various sources. The goal of the TMDL is to restrict loadings so that ambient pollution concentrations will decrease to levels that no longer contribute to impairment.

Ambient water TIEs proceed in the same manner as effluent TIEs using the USEPA Phase I, II, and III protocols already described (USEPA 1991, 1992, 1993a,b, 1996). Once toxicity is observed in an ambient water sample, the cause can be investigated with the TIE process. In contrast to effluent samples that usually provide a consistent toxicity signal, ambient waters tend to be influenced by non-point sources of toxicity and could have intermittent toxicity caused by pulses of contaminants. Generally, if toxicity is suspected in ambient water, sufficient volume is collected for initial toxicity testing as well as amounts adequate for the entire TIE process.

Sediments act as a sink for many contaminants and are often viewed as integrators of anthropogenic impacts (USEPA 2004). Because many contaminants, particularly hydrophobic chemicals, can persist in sediments, the toxicity signal is generally more persistent. Sediment TIEs are conducted using solid-phase and interstitial water treatments. As with the ambient water samples, interstitial water TIEs follow the USEPA protocols with minor modifications for volume. These modifications, along with a suite of solid-phase treatment methods, are described in an upcoming USEPA sediment TIE guidance document (USEPA, 2007), and in Anderson et al. (in press).

After sediment toxicity is observed, a decision matrix guides the TIE process as described by Anderson et al. (in press). The process begins with an initial assessment of the magnitude of toxicity using solid-phase and interstitial water dilution series tests. Solid-phase dilutions may be necessary when complete mortality is observed in the original sample to allow improved resolution between baseline and TIE treatments. Once the magnitudes of toxicity are determined, a single sediment concentration is selected for the solid-phase TIE, and a dilution series is selected for the interstitial water TIE. In this approach, solid-phase TIEs are generally conducted on a single concentration of sediment for efficiency but can be conducted on multiple concentrations if resources allow. The solid-phase concentration is chosen to provide the best resolution among the TIE treatments. Interstitial water TIEs are conducted with a dilution series that brackets the LC_{50} from the initial test. In situations where one sediment matrix produces insufficient toxicity to allow resolution of differences between the TIE treatment and baseline, the TIE proceeds using the matrix with the highest magnitude of toxicity.

4.4 Identification of Pyrethroid-Associated Toxicity with TIEs

Current TIE methodology can characterize and, in some cases, identify pyrethroid toxicity (Anderson et al., in press). However, these methods are still imprecise, and further work is needed to develop specific methods for pyrethroid identification. Even the most robust of these methods are limited in their ability to quantify toxicity contributed by individual pyrethroids when these pesticides occur in mixtures (e.g., permethrin vs. deltamethrin). Previous research has shown that a weight-of-evidence approach is effective at identifying sediment toxicity caused by pyrethroids. TIE evidence includes solid-phase extraction and elution of interstitial water samples, addition of carbonaceous resin to whole-sediment samples, and solvent elution of resins to recover sorbed pesticides. Lines of evidence for pyrethroid toxicity are provided when results from these procedures are combined with detailed chemical analyses and information from the standard TIE treatments.

4.4.1 Use of Temperature Differential

One potential method for characterizing pyrethroid-associated toxicity involves the use of temperature. Pyrethroids have been reported to exhibit a negative temperature coefficient (toxicity increasing with decreasing ambient temperature) in varying degrees depending upon pyrethroid and species (Pap 2003). This effect is thought to be partly caused by reduced metabolism of the parent pyrethroid to less-toxic metabolites (Ware and Whitacre 2004). In support of this theory, increased toxicity of sediments and/or interstitial waters containing pyrethroids has been observed in *H. azteca* tests conducted at 15°C compared with concurrent tests conducted at 23°C (Anderson et al. 2006b; Weston 2006; Phillips et al. 2007; Anderson et al., in press-b). However, we are unaware of any published reports that utilized reduced temperature toxicity tests as part of TIEs aimed at identifying pyrethroid-caused toxicity in ambient freshwaters. Therefore, we performed studies to determine if reduced temperature toxicity tests would be useful in a TIE context. These studies were aimed at the identification of pyrethroid-caused toxicity in two commonly used freshwater test species, fathead minnows (*Pimephales promelas*) and *C. dubia*, using permethrin, bifenthrin, cypermethrin, esfenvalerate, and cyfluthrin at 15°C and 25°C. The 48-hr acute *C. dubia* and 96-hr acute fathead minnow bioassays were conducted in accordance with the U.S. Environmental Protection Agency (USEPA) 5th edition protocol (USEPA 2002), as described in Tables 4 and 5.

Table 4 Effect of reduced temperature on acute toxicity of pyrethroids to *Ceriodaphnia dubia*[a]

	25°C	15°C	
Pyrethroid (type)	LC_{50}[b] (µg/L) (SD)	LC_{50} (µg/L) (SD)	Toxicity ratio[c]
Permethrin (I)	0.21 (0)	0.078 (0)	2.7
Bifenthrin (I)	0.074 (0.004)	0.045 (0.001)	1.6
Cypermethrin (II)	0.22 (0.005)	0.27 (0.032)	0.8
Esfenvalerate (II)	0.14 (0.022)	0.089 (0.007)	1.6
Cyfluthrin (II)	0.21 (0)	0.12 (0.014)	1.8

[a]*C. dubia* acute toxicity tests were conducted with neonates <24 hr old collected within an 8-hr period from in-house cultures. Pyrethroid analytical standards (≥99% pure) obtained from AccuStandard (New Haven, CT) were diluted in HPLC-grade methanol. Each test consisted of 5–6 concentrations of the test material with 4 replicates of 5 neonates each. Test chambers were 20-mL glass scintillation vials containing 18 mL test solutions. Dilution water was reverse osmosis- and granular carbon-treated well water amended with dry salts to attain USEPA moderately hard specifications (EPAMH). Test duration was 48 hr; test solutions were not renewed and test organisms were not fed during the test. Tests were conducted in environmental chambers at 15° ± 2°C and 25° ± 2°C. Photoperiod was 16 hr light:8 hr dark. Mortality was noted daily. LC_{50} values were calculated using ToxCalc (Tidepool Scientific, McKinleyville, CA).
[b]LC_{50} values are nominal concentrations.
[c]Toxicity ratio = LC_{50} at 25°C/LC_{50} at 15°C.

Table 5 Effect of reduced temperature on acute toxicity of pyrethroids to larval fathead minnows (*Pimephales promelas*)[a]

	25°C	15°C	
Pyrethroid (type)	LC_{50}[b] (µg/L) (SD)	LC_{50} (µg/L) (SD)	Toxicity ratio[c]
Permethrin (I)	0.86 (0.03)	1.97 (0.54)	0.4
Bifenthrin (I)	0.52 (0.03)	0.49 (0.04)	1.1
Cypermethrin (II)	2.52 (0.18)	4.00 (0.08)	0.6
Esfenvalerate (II)	0.49 (0.04)	0.38 (0.04)	1.3
Cyfluthrin (II)	0.79 (0.02)	1.37 (0.23)	0.6

[a]Fathead minnows were obtained from Aquatox, Inc. (Hot Springs, AK), and were maintained in USEPA moderately hard (EPAMH) water until tested at 7 d old. Pyrethroid analytical standards (≥99% pure) obtained from AccuStandard (New Haven, CT) were diluted in HPLC-grade methanol. Each test incorporated 5–6 dilutions of the test material using 2 replicates of 10 fish each in 400-mL glass beakers containing 250 mL test solutions. Dilution water was EPAMH laboratory water. Test duration was 96 hr, and test solutions were renewed at 48 hr. Fish were fed *Artemia nauplii* 4 hr before sample renewal at 24 hr. Tests were conducted in an environmental chamber at 15° ± 2°C and 25° ± 2°C. Photoperiod was 16 hr light:8 hr dark. Mortality was noted daily. LC_{50} values were calculated using ToxCalc (Tidepool Scientific, McKinleyville, CA).
[b]LC_{50} values are nominal.
[c]Toxicity ratio = LC_{50} at 25°C/LC_{50} at 15°C.

Results of the temperature study were mixed. The LC_{50} values for *C. dubia* ranged from 0.074–0.22 µg/L at 25°C to 0.045–0.27 µg/L at 15°C (see Table 4). Bifenthrin was most toxic and cypermethrin the least toxic, regardless of test temperature. The toxicity ratios (TRs; LC_{50} at 25°C/LC_{50} at 15°C) were calculated for each of the pyrethroids. If the reduced temperature resulted in increased toxicity, the TRs would be greater than unity. The value for cypermethrin was less than

1.0 whereas the TR values for the other four pyrethroids ranged from 1.6 to 2.7. Thus, with the exception of cypermethrin, reduced temperature resulted in approximately a two- to threefold increase in toxicity and would therefore be useful for identification of pyrethroid-caused toxicity to *C. dubia*.

The results with fathead minnows were quite different from the *C. dubia* studies. At 25°C, LC_{50} values ranged from 0.49 to 2.52 μg/L, compared to 0.49–4.00 μg/L at 15°C. As with the *C. dubia* tests, bifenthrin was the most toxic and cypermethrin the least toxic regardless of test temperature. Table 5 shows that the TR values ranged from 0.4 to 1.1 for Type I pyrethroids and from 0.6 to 1.3 for Type II pyrethroids. In these tests, none of the pyrethroids tested exhibited significantly increased toxicity at 15°C compared to 25°C and, interestingly, three of the pyrethroids (permethrin, cypermethrin, and cyfluthrin) were noticeably less toxic at the lower test temperature. These data indicate that reduced temperature would not be a useful tool for identification of pyrethroid-caused toxicity to fathead minnows. In contrast, for *C. dubia*, reduced temperature toxicity tests would be a useful TIE procedure, adding to the weight-of-evidence for identification of pyrethroid-caused toxicity in aqueous samples.

4.4.2 Use of Piperonyl Butoxide (PBO)

In addition to the use of SPE columns for water samples and resin for sediment samples to characterize the cause of toxicity as an organic, the addition of PBO can characterize pyrethroid-associated toxicity by synergizing the toxicity signal (Kakko et al. 2000; Ameg and Weston, 2007). Certain compounds (e.g., OPs) must be metabolically activated by the test organism before they can exert their toxic effect (Casida and Quistad 2004). Many of these activation reactions consist of oxidative metabolism by a group of enzymes collectively known as mixed-function oxidases (MFOs), of which the heme-protein cytochrome P450s are a subset (Hodgson 1982) (see Fig. 5). Compounds such as PBO, a synthetic methylenedioxy phenyl compound, bind to and block the catalytic activity of some MFOs, preventing the toxicity of metabolically activated OP insecticides, such as diazinon, chlorpyrifos, malathion, parathion, and fenthion (Hamm et al. 2001). Thus, when a nontoxic level of PBO is added to test samples containing one or more of these OPs, the toxicity is greatly reduced or completely blocked. The use of PBO to identify toxicity caused by metabolically activated OP insecticides has been previously described (Ankley et al. 1991) and incorporated in published USEPA TIE manuals (USEPA 1991, 1993b). Conversely, PBO synergizes the toxicity of pyrethroid insecticides by blocking MFO-mediated metabolism of these chemicals (Casida 1970). Thus, PBO addition to samples containing one or more pyrethroids increases and/or prolongs the toxic effect (Casida and Quistad 1995). This dual action of PBO can lead to a confounding signal in TIE testing of samples that contain both OPs and pyrethroids. Alternatively, the action of PBO can be a useful tool to identify the presence/absence of pyrethroids and metabolically activated OPs in aqueous samples. Accordingly, it is important that the use of PBO in TIE testing is performed with the correct controls to enable interpretation of the results. Several studies have successfully utilized PBO addition to build a weight-of-evidence for pyrethroid toxicity in ambient water and sediment (Anderson et al. 2006b; Phillips et al. 2006, 2007; Amweg and Weston 2007; Anderson et al., in press).

4.4.3 Use of Carboxylesterase Activity

A novel method for pyrethroid detection involving the addition of carboxylesterase during the TIE process was proposed by Wheelock and coworkers (2004, 2006). As discussed previously, carboxylesterases are very efficient at hydrolyzing pyrethroids to their corresponding acid and alcohol, which generally significantly reduces observed toxicity. Carboxylesterases rapidly degrade both Type I and Type II pyrethroids (see Fig. 1). This class of enzymes has been demonstrated to be effective in reducing pyrethroid-associated toxicity in both mammals and insects (Abernathy and Casida 1973) and is a logical tool for removing the toxicity of ester-containing pyrethroids in TIE samples. Previous work identified this enzyme as a good target for identifying pyrethroid-associated toxicity in aquatic samples (Denton et al. 2003). Carboxylesterase addition to aqueous samples, or water overlying sediment in solid-phase exposures, provides evidence of toxicity somewhere between the Phase I TIE (characterization) and Phase II TIE (identification). If toxicity is reduced after enzyme addition, the probable cause of toxicity is attributed to a pyrethroid. However, this test does not identify the specific pyrethroid. Previous work with carboxylesterase addition to identify/remove pyrethroid-associated toxicity validated the concept with *C. dubia* using several pyrethroids, including permethrin, bifenthrin, cypermethrin, cyfluthrin, lambda-cyhalothrin, and esfenvalerate (Wheelock et al. 2004) (Table 6). Additional studies with *H. azteca* demonstrated that nonpyrethroid toxicants (e.g., DDT, cadmium, and chlorpyrifos) were not significantly affected by carboxylesterase addition (Weston and Amweg 2007).

The carboxylesterase preparation method is thoroughly described in Wheelock et al. (2004, 2006) and only briefly here. Carboxylesterase is commercially available and can be purchased as a liquid (ammonium sulfate suspension, 3.2 M, pH 8.0) or a crude lyophilized powder preparation. Each preparation has its individual advantages and disadvantages, which have been evaluated elsewhere (Wheelock et al. 2006). Some researchers have a preference for the liquid preparation for ease of pipetting, whereas others believe that the powder is simpler to handle. In addition, there are specific toxicity issues associated with each preparation because the ammonia concentration of the liquid preparation is extremely high. The commercial carboxylesterases are available as preparations from porcine or rabbit liver. For reasons of cost, all studies to date have used the porcine preparations. Standardized methods for sediment testing with *H. azteca* use the lyophilized enzyme. Before test initiation, the lyophilized enzyme powder is added to Nanopure water to prepare a stock solution that can be added directly to the sample. The amount of enzyme added depends on sample volume and the tolerance of the test organism. It is important that toxicity of the enzyme preparation to the organism is evaluated

Table 6 Effect of carboxylesterase addition upon pyrethroid- and nonpyrethroid-associated toxicity to *Ceriodaphnia dubia* and *Hyalella azteca*

Treatment	Conc.[a] (ng/L)	Esterase (–)	Esterase (+)
	C. dubia 48-hr % mortality[b]		
Control[c]	0	0	0
Permethrin	600	100	0
Bifenthrin	660	100	0
Cypermethrin	1450	100	0
Esfenvalerate	700	100	0
Cyfluthrin	560	100	0
λ-Cyhalothrin	600	100	0
Diazinon	760	100	100
Chlorpyrifos	160	100	100
	H. azteca 10-d % mortality[d]		
Cadmium	—[e]	17	10
DDT	—	13	7
Chlorpyrifos	—	13	7

[a]Conc. indicates the concentration of insecticide at which the samples were spiked. All values are given as nominal water concentrations. The values in parentheses are toxic units (TUs), which are defined as $100/EC_{50}$ (the concentration at which 50% of the population exhibit an effect).
[b]The mean mortality in two replicates of five neonate *C. dubia*. The standard deviations for all samples were zero. Esterase (–) samples contain no added enzyme. Esterase (+) samples contain commercial porcine carboxylesterase spiked at 2.5×10^{-3} U/ml (purchased from Sigma Chemical). Data are from Wheelock et al. (2004).
[c]Controls were performed with and without the addition of the esterase to account for potential esterase-associated toxicity.
[d]Data are from Weston and Amweg (2007).
[e]Toxicant concentrations are not given as studies were performed with a range of values to determine LC_{50}s in the presence and absence of esterase. The LC_{50}s were measured for cadmium (control: 71.9 mg/kg, esterase 267 mg/kg), chlorpyrifos [control: 2.96 μg/g organic carbon (oc); esterase 4.20 μg/g oc], and DDT (control: 147 μg/g oc, esterase 174 μg/g oc).

because toxicity has been observed at increased concentrations of the enzyme (most likely from the presence of ammonia). The carboxylesterase is added based upon units of enzyme activity, which are defined by the supplier. In this case, one unit (U) is defined as the amount of enzyme required to hydrolyze 1.0 μmole of ethyl butyrate to butyric acid and ethanol per minute at pH 8.0 at 25°C. We have developed a dilution nomenclature based upon enzyme activity, wherein "X" units of enzyme activity equals 0.0025 U/mL sample; therefore, at 500 X, 1.25 U are added per milliliter (mL) sample. Typically, *H. azteca* can tolerate an enzyme strength of 500 X without showing toxicity. Enzyme activity is unique for each lot purchased and should be defined on a case-by-case basis (Wheelock et al. 2006). The enzyme should be added to the water overlying the sediment on the day of test initiation at least 6 hr before the organism addition. This interval provides the enzyme with sufficient time to hydrolyze pyrethroids present in the sample. Specific activity assays performed with the porcine esterase showed that hydrolysis was pyrethroid structure dependent, with Type II pyrethroids hydrolyzed more slowly than those of Type I. However, given the rate of hydrolysis (160 nmol/min/mg protein for permethrin vs. 23 nmol/min/mg protein for cyfluthrin), it is likely that the majority of pyrethroid present in the sample will be hydrolyzed within 6 hr. If the samples are renewed, for example, during a 96-hr testing cycle, then the esterase is also added with the new water sample.

One concern in the use of the esterase in toxicity assays is the specificity of the observed reduction in toxicity. As already discussed, pyrethroids are extremely hydrophobic and may adsorb to the hydrophobic enzyme surface without undergoing hydrolysis and subsequent detoxification. In addition, it is possible that other hydrophobic contaminants in the system could also undergo nonspecific absorption interactions with the enzyme, thereby eliminating the selectivity of the method. It is therefore suggested that a separate test be performed to control for nonspecific reductions in observed toxicity. To accomplish this, BSA can be added to a separate set of sample replicates. BSA is a noncatalytically active protein that is approximately the same size as carboxylesterase (~67 kDa for BSA vs. ~65 kDa for carboxylesterase). The BSA stock solution is prepared using the same mass of protein and volume of Nanopure water as used with the enzyme, and the same volume of stock solution is added to the test solution. Accordingly, if a reduction in toxicity is observed in the BSA samples, it can be attributed to nonspecific adsorption effects instead of specific catalytic reductions in pyrethroid levels. However, it should be cautioned that carboxylesterase activity has occasionally been found in commercial BSA preparations. It is recommended that the enzyme be added daily to the water column exposures and to the water overlying the sediment to ensure complete hydrolysis of pyrethroids. This procedure may also be used in a combined treatment with the addition of PBO to determine if both OPs and pyrethroids are contributing to the observed toxicity; however, results can be difficult to interpret because of the counterindicative effects.

5 Applications of Carboxylesterase Activity in TIEs

The following examples of the use of carboxylesterase additions in freshwater and marine ambient waters, sediment, and interstitial waters are summarized in Table 7.

5.1 Water Column

Water samples collected from the New River (CA, USA), where it crosses the border with Mexico, contain complex mixtures of contaminants from agriculture, and municipal, and industrial effluents. A series of TIEs implicated several classes of contaminants as the cause of observed toxicity (Phillips et al. 2007), but the most recent TIEs using carboxylesterase, along with SPE and subsequent elution of the SPE column, identified cypermethrin as the actual cause of toxicity. At the time of analysis, the laboratory detection limit for cypermethrin was well above the toxicity threshold of the test organism, but after the enzyme characterized the cause of

Table 7 Summary of reported water and sediment (solid phase and interstitial water) TIEs that have utilized carboxylesterase addition

Location and reference	Carboxylesterase success	Additional TIE evidence	Chemical analysis
Water			
New River, CA, USA (Phillips et al. 2007)	Reduced toxicity	Return of toxicity with Phase II elution of SPE column	Detection of toxic concentration of cypermethrin in SPE eluate
Central Valley, CA, USA (AQUA-Science, Inc.)	Reduced toxicity	Increased toxicity with PBO addition	None
Sediment			
Solid phase:			
Orcutt Creek, CA, USA (Phillips et al. 2006)	Reduced toxicity	Increased toxicity with PBO addition	Concentrations of pyrethroids in sediment were below LC_{50} values
Westley Wasteway, CA, USA (Anderson et al., in press)	Reduced toxicity	Some evidence in concurrent interstitial water TIE	Detection of toxic concentration of lambda-cyhalothrin in sediment and resin eluate
Agricultural Tailwater Pond SV-03, Salinas, CA, USA (Anderson et al., in press)	Reduced toxicity	Some evidence in concurrent interstitial water TIE	Detection of toxic concentrations of cypermethrin and lambda-cyhalothrin in sediment
Grayson Drain, CA, USA (MPSL 2006b)	Reduced toxicity, but also had reduction with BSA addition	None	No pyrethroids detected in sediment
Del Puerto Creek, CA, USA (MPSL 2006a)	Reduced toxicity	Some evidence in concurrent interstitial water TIE	Detection of toxic concentration of bifenthrin in sediment and resin eluate
Alisal Creek, CA, USA (Anderson et al. 2007b)	Reduced toxicity	Some evidence in concurrent interstitial water TIE	Detection of toxic concentration of pyrethroids in sediment and resin eluate
Indiana Harbor, IN, USA (Anderson et al. 2007b)	Reduced toxicity, but also had reduction with BSA addition	None	No pyrethroids detected in sediment
San Diego Creek, CA, USA (Anderson et al. 2007b)	Reduced toxicity	Increased toxicity with PBO addition	Detection of toxic concentrations of bifenthrin and permethrin in resin eluates
Upper Newport Harbor, CA, USA (Anderson et al. 2007b)	None	None	Detection of toxic concentration of bifenthrin in sediment

Interstitial water:			
Santa Maria River, CA, USA (Anderson et al. 2006b)	Reduced toxicity	Increased toxicity with lower temperature and PBO addition	Detection of toxic concentration of lambda-cyhalothrin in sediment
Westley Wasteway, CA, USA	Reduced toxicity	Increased toxicity with PBO addition	Detection of toxic concentration of lambda-cyhalothrin in sediment
Agricultural Tailwater Pond SV-03, Salinas, CA, USA (Anderson et al., in press)	Reduced toxicity	Increased toxicity with PBO addition	Detection of toxic concentrations of cypermethrin and lambda-cyhalothrin in sediment
Del Puerto Creek, CA, USA (MPSL 2006a)	Reduced toxicity	Increased toxicity with PBO addition	Detection of toxic concentration of bifenthrin in sediment
Alisal Creek, CA, USA (Anderson et al. 2007b)	Reduced toxicity	Increased toxicity with PBO addition	Detection of toxic concentration of pyrethroids in interstitial water
Indiana Harbor, IN, USA (Anderson et al. 2007b)	Reduced toxicity, but also had reduction with BSA addition	None	No pyrethroids detected in interstitial water
Upper Newport Harbor, CA, USA (Anderson et al. 2007b)	Reduced toxicity	Increased toxicity with PBO addition	Detection of toxic concentration of bifenthrin in sediment

PBO, piperonyl butoxide; BSA, bovine serum albumin.

toxicity as a pyrethroid, the solvent elution of the SPE was concentrated and analyzed directly. The results of this direct analysis demonstrated the presence of cypermethrin at a toxic concentration.

In another ambient water sample collected from the central valley of California, addition of PBO synergized toxicity as much as 30 fold (AQUA-Science, Inc., personal communication). The addition of carboxylesterase completely removed toxicity. The enzyme was also used in a stacked treatment with PBO to determine if OPs contributed to observed toxicity. PBO was added several hours after the enzyme, but no mortality was observed, suggesting pyrethroids were the sole cause of toxicity. However, the sample was not chemically analyzed for pyrethroids, so the authors were not able to positively identify the cause of toxicity.

5.2 Sediment

Because pyrethroids are hydrophobic, they deposit in sediments, and there is growing evidence of their presence in sediments associated with agricultural and urban areas (Weston et al. 2004, 2005; Gan et al. 2005; Amweg et al. 2006). Although pyrethroids have been attributed to observed toxicity in several studies based on concentration alone (Weston et al. 2005), few comprehensive sediment TIE studies have been conducted. Through the use of carboxylesterase, several of these studies have demonstrated that pyrethroids contributed to toxicity.

Results from Anderson et al. (2006b) suggested that pyrethroids were contributing to interstitial water toxicity in a sample from the lower Santa Maria River (CA, USA), a watershed heavily impacted by agriculture. The results of the carboxylesterase treatment in this TIE were inconclusive because of an elevated concentration of chlorpyrifos. It was assumed that the addition of enzyme reduced pyrethroid-associated toxicity, but no reduction of toxicity was observed because of the presence of a toxic concentration of chlorpyrifos. However, addition of PBO did not reduce chlorpyrifos toxicity because of the presence of a pyrethroid and resulting synergism. Although interstitial water chemistry was not conducted, the concentration of lambda-cyhalothrin in the sediment exceeded the solid-phase LC_{50} for the test organism. In a companion paper, the TIE results of Phillips et al. (2006) suggested that pyrethroids contributed to toxicity in sediment from Orcutt Creek, a tributary of the Santa Maria River. This study used carboxylesterase in the overlying water of a standard solid-phase toxicity test to determine if pyrethroids were contributing to the toxicity of *H. azteca*. Carboxylesterase addition increased organism survival from 15% to 60%.

Anderson et al. (in press) provide an up-to-date review of current sediment TIE methods for pyrethroids. They also present two sediment toxicity case studies that illustrate the successful use of carboxylesterase. Both case study sediments are from agricultural drainages in California and include the application of carboxylesterase in both solid-phase and interstitial water exposures. In the first case study, toxicity in the solid-phase exposures from station WWNCR (Westley Wasteway) was not reduced with the addition of an organic-binding resin, but the addition of the enzyme increased survival from 0% to 48%. Addition of carboxylesterase to interstitial water decreased toxicity from 5.6 toxic units (TUs; a TU is defined as $100/EC_{50}$, where EC_{50} is the concentration at which 50% of the population exhibits an effect) in the untreated baseline sample to 1.7 TUs. Addition of BSA to water overlying sediment and interstitial water did not significantly affect toxicity relative to the baseline sample. Chemical analysis of the sediment detected 14 μg/g organic carbon of lambda-cyhalothrin [approximately 30 TUs based on the LC_{50} of Amweg et al. (2005)]. Addition of the carbonaceous resin Ambersorb did not reduce toxicity, but when the resin was recovered from the sediment and eluted with acetone, the eluate was toxic. Analysis of the eluate detected 1279 ng/L lambda-cyhalothrin. When combined with evidence of reduced toxicity from esterase addition, these results provide compelling evidence that this pyrethroid was the primary cause of toxicity.

In the second case study presented by Anderson et al. (in press), solid-phase and interstitial water TIEs were conducted on sediment collected from an agricultural tailwater pond. Addition of carboxylesterase increased survival of *H. azteca* in both the water overlying the sediment and the interstitial water. The addition of BSA did not reduce toxicity, supporting pyrethroids as the cause of toxicity. Analysis of the sediment identified toxic concentrations of cypermethrin and lambda-cyhalothrin (18 TUs, and 2 TUs, respectively).

Carboxylesterase addition was also used successfully in two studies conducted by the University of California, Davis, Marine Pollution Studies Laboratory (MPSL 2006a,b). Sediment from two agricultural drainages, Del Puerto Creek and Grayson Drain, caused high mortality in initial toxicity tests. Addition of the enzyme to overlying water in solid-phase exposures with Grayson Drain sediment reduced toxicity, but additions of BSA also completely removed toxicity. Analysis of the sediment did not detect any pyrethroids, but concentrations of DDT metabolites in the resin eluate exceeded published LC_{50} values for *H. azteca.* The toxicity of Del Puerto Creek sediment was also reduced by the addition of carboxylesterase to the water overlying the sediment, and BSA addition did not reduce toxicity. A TIE with interstitial water was also conducted with similar results: addition of the enzyme reduced toxicity and BSA addition did not. Analysis of Del Puerto Creek sediment and the resin eluate detected toxic concentrations of bifenthrin.

As part of a larger study to evaluate current and emerging sediment TIE methods, carboxylesterase was used as a standard treatment in six ambient sample TIEs utilizing solid-phase and interstitial water matrices (Anderson et al. 2007b). Carboxylesterase was evaluated using three freshwater samples and one marine sample. The toxicity of sediment and interstitial water from Alisal Creek (CA, USA) was reduced by the addition of the enzyme. A toxic concentration of lambda-cyhalothrin was detected in the sediment, and additional pyrethroids were identified in the interstitial water and in the solvent eluate of the solid-phase resin. Two additional stations from southern California included enzyme additions in TIEs: San Diego Creek and Upper Newport Harbor. Both stations are heavily influenced by urban inputs and a small percentage of agriculture, and San Diego Creek is a tributary to Newport Bay. Addition of carboxylesterase reduced the toxicity of San Diego Creek sediment, but the enzyme treatment was apparently saturated by the toxicity of Upper Newport sediment. Addition of the enzyme to Upper Newport interstitial water reduced toxicity. Bifenthrin and permethrin were detected in resin eluates from the solid-phase treatments from both samples, and in the sediment from Upper Newport, indicating that pyrethroids contributed to the toxicity of these samples.

Weston and Ameg (2007) formally evaluated the solid-phase TIE treatment by adding carboxylesterase to the overlying water of spiked and ambient sediments in tests with *H. azteca*. Addition of enzyme to the water overlying the sediment successfully reduced the toxicity of bifenthrin-spiked sediment. The enzyme also reduced the toxicity of cadmium-spiked sediment but did not reduce the toxicity of sediments spiked with DDT or chlorpyrifos. The authors also evaluated the enzyme by adding it to the overlying water of 12 ambient samples containing known toxic concentrations of various pyrethroids. The addition of carboxylesterase significantly increased survival in all the samples, but in 3 of these samples the increase was not significantly different from that observed following the addition of BSA.

5.3 Method Limitations

As discussed, carboxylesterase addition can help identify pyrethroid-associated toxicity in water and sediment samples. However, there are limitations with the current method that should be addressed to improve this tool. One of the main limitations is that the carboxylesterase preparation, either liquid or lyophilized powder, is heterogeneous. The commercial product is prepared from pig or rabbit livers and undergoes a simple acetone precipitation before sale. Subsequently, a large amount of undefined noncatalytically active protein is added to the test system, greatly increasing the likelihood of nonspecific interactions with hydrophobic contaminants. To overcome this problem, it would be ideal to use a recombinant protein selected for a high specific activity for pyrethroid hydrolysis in these assays, along the lines of that identified by Stok et al. (2004a). If available, such a preparation would provide high enzymatic activity while minimizing the concentration of nonspecific proteins in the assay. However, making this preparation would require protein expression and purification on a commercial scale to obtain sufficient quantities of purified protein.

Another problem with the commercial preparation is the reproducibility of the preparation. Experiments in which small amounts of the liquid esterase preparation were aliquotted into 20-mL borosilicate glass containers exhibited high variability (21% relative standard deviation; RSD) in contrast to tests with larger amounts of esterase (10% RSD) (Wheelock et al. 2006). In addition, vial-to-vial variability was ~11% among five different bottles examined. Another issue is variability among lots. Each vial of carboxylesterase has a certain level of activity, which is defined by the supplier/manufacturer. However, this activity can vary greatly among lots. This variability requires adjusting the total amount of protein or the total amount of enzyme activity added with each new lot purchased. Although none of these points prevents the use of the enzyme in TIEs, identifying an alternative enzyme source could greatly increase the utility of this method. In particular, different aquatic testing laboratories could more easily compare results if they used an identical enzyme preparation.

A number of other concerns regarding the esterase preparation were identified by Weston and Amweg (2007), who performed an extensive evaluation of the use of carboxylesterase activity in whole-sediment testing. One major concern is the physical amount of material added to the test system. The mass of carboxylesterase used in this method has been demonstrated to be sufficient to decrease the oxygen levels in testing systems, with dissolved oxygen dropping by more than 50% within 24 hr of test initiation. Studies performed with mercuric chloride addition eliminated the oxygen depletion, strongly suggesting that the observed reductions were the result of increased microbial activity. These observations were further supported by control studies with BSA, which also caused a drop in oxygen levels (albeit slower than that following esterase addition). However, similar concerns have not been noted by other researchers.

One potential way to increase the efficacy of the assay is to increase the incubation time of the esterase with the sample to be tested before organism addition. Current methods use an 1- to 6-hr incubation; however, it may be better to incubate for 12–24 hr to ensure significant hydrolysis of pyrethroids before organism addition. Assays could be formatted such that carboxylesterase was introduced to the sample the evening before test initiation/organism addition. The increased incubation time may increase the efficiency of pyrethroid removal/hydrolysis from the testing solution.

6 Summary

This review has examined a number of issues surrounding the use of carboxylesterase activity in environmental monitoring. It is clear that carboxylesterases are important enzymes that deserve increased study. This class of enzymes appears to have promise for employment in environmental monitoring with a number of organisms and testing scenarios, and it is appropriate for inclusion in standard monitoring assays. Given the ease of most activity assays, it is logical to report carboxylesterase activity levels as well as other esterases (e.g., acetylcholinesterase). Although it is still unclear as to whether acetylcholinesterase or carboxylesterase is the most "appropriate" biomarker, there are sufficient data to suggest that at the very least further studies should be performed with carboxylesterases. Most likely, data will show that it is optimal to measure activity for both enzymes whenever possible. Acetylcholinesterase has the distinct advantage of a clear biological function, whereas the endogenous role of carboxylesterases is still unclear. However, a combination of activity measurements for the two enzyme systems will provide a much more detailed picture of organism health and insecticide exposure. The main outstanding issues are the choice of substrate for activity assays and which tissues/organisms are most appropriate for monitoring studies. Substrate choice is very important, because carboxylesterase activity consists of multiple isozymes that most likely fluctuate on an organism- and tissue-specific basis. It is therefore difficult to compare work in one organism with a specific substrate with work performed in a different organism with a different substrate. An attempt should therefore be made to standardize the method. The most logical choice is PNPA (*p*-nitrophenyl acetate), as this substrate is commercially available, requires inexpensive optics for assay measurements, and has been used extensively in the literature. However, none of these beneficial properties indicates that the substrate is an appropriate surrogate for a specific compound, e.g., pyrethroid-hydrolyzing activity. It will most likely be necessary to have more specific surrogate substrates for use in assays that require information on the ability to detoxify/hydrolyze specific environmental contaminants.

The use of carboxylesterase activity in TIE protocols appears to have excellent promise, but there are further technical issues that should be addressed to increase the utility of the method. The main concerns include the large amount of nonspecific protein added to the testing system, which can lead to undesirable side effects including nonspecific reductions in observed toxicity, decrease in dissolved oxygen content, and organism growth. It is probable that these issues can be resolved with further assay development. The ideal solution would be to have a commercial recombinant carboxylesterase that possessed elevated pyrethroid-hydrolysis activity and which was readily available, homogeneous, and inexpensive. The availability of such an enzyme would address nearly all the current method shortcomings. Such a preparation would be extremely useful for the aquatic toxicology community. Further work should focus on screening available esterases for stability, cost, and activity on pyrethroids, with specific focus on esterases capable of distinguishing type I from type II pyrethroids. It would also be beneficial to identify esterases that are not sensitive to OP insecticides. Many esterases and lipases are available as sets to test chemical reactions for green chemistry, enabling large-scale screening. Other potential approaches to increase the utility of the enzyme include derivatization with polyethylene glycol (PEG) or cyanuric acid chloride to increase stability and reduce microbial degradation. It is also possible that the enzyme could be formulated in a sol gel preparation to increase stability. It is likely that the use of carboxylesterase addition will increase for applications in sediment TIEs.

Carboxylesterases are an interesting and useful enzyme family that deserves further study for applications in environmental monitoring as well as to increase our understanding of the fundamental biological role(s) of these enzymes. There are, of course, other enzymes that show high esterase activity on pyrethroids but are not technically carboxylesterases in the α/β-hydrolase fold protein family. These enzymes should also be examined for use in TIE protocols and "esterase" arrays as well as for general applications in environmental monitoring. One can envision the creation of a standardized screen of enzymes with esterase activity to (1) identify environmental contaminants, (2) estimate the potential toxic effects of new compounds on a range of organisms, and (3) monitor organism exposure to agrochemicals (and potentially other contaminants). This approach would provide a multibiomarker integrative assessment of esterase-inhibiting potential of a compound or mixture. In conclusion, much is still unknown about this enzyme family, indicating that this area is still wide open to researchers interested in the applications of carboxylesterase activity as well as basic biological questions into the nature of enzyme activity and the endogenous role of the enzyme.

Acknowledgments We thank the staff of the UC Davis Marine Pollution Studies Laboratory, Sara Clark, Jennifer Vorhees, Katie Siegler, and Jason Flynn. We thank Peter Buchwald for generation of the carboxylesterase structure and Juan Sanchez-Hernandez for critical reading of the manuscript. C.E.W. was supported by an EU Sixth Framework Programme (FP6) Marie Curie International Incoming Fellowship (IIF). This work was supported in part by NIEHS Grant R37 ES02710, NIEHS Superfund Basic Research Program Grant P42 ES04699, and USDA Grant 2007-35607-17830.

References

Abdel-Aal YAI, Hammock BD (1986) Transition state analogs as ligands for affinity purification of juvenile hormone esterase. Science 233:1073–1075.

Abernathy CO, Casida JE (1973) Pyrethroid insecticides: esterase cleavage in relation to selective toxicity. Science 179:1235–1236.

Adams MS, Stauber JL (2004) Development of a whole-sediment toxicity test using a benthic marine microalga. Environ Toxicol Chem 23:1957–1968.

Ahmad M, Arif MI, Ahmad Z, Denholm I (2002) Cotton whitefly (*Bemisia tabaci*) resistance to organophosphate and pyrethroid insecticides in Pakistan. Pestic Manag Sci 58:203–208.

Ahmad S, Forgash AJ (1976a) Nonoxidative enzymes in the metabolism of insecticides. Drug Metab Rev 5:141–164.

Ahmad S, Forgash AJ (1976b) Nonoxidative enzymes in the metabolism of insecticides. Ann Clin Biochem 13:141–164.

Al-Ghais SM, Ahmad S, Ali B (2000) Differential inhibition of xenobiotic-metabolizing carboxylesterases by organotins in marine fish. Ecotoxicol Environ Saf 46:258–264.

Aldridge WN (1953a) Serum esterases 1. Two types of esterase (A and B) hydrolysing *p*-nitrophenyl acetate, propionate and butyrate, and a method for their determination. Biochem J 53:110–117.

Aldridge WN (1953b) Serum esterases 2. An enzyme hydrolysing diethyl *p*-nitrophenyl phosphate (E600) and its identity with the A-esterase of mammalian sera. Biochem J 53:117–124.

Aldridge WN (1990) An assessment of the toxicological properties of pyrethroids and their neurotoxicity. Crit Rev Toxicol 21:89–104.

Aldridge WN (1993) The esterases: perspectives and problems. Chem Biol Interact 87:5–13.

Amweg EL, Weston DP (2007) Whole sediment toxicity identification evaluation tools for pyrethroid insecticides: I. Piperonyl butoxide addition. Environ Toxicol Chem 26:2389–2396.

Amweg EL, Weston DP, Ureda NM (2005) Use and toxicity of pyrethroid pesticides in the Central Valley, CA, U.S. Environ Toxicol Chem 24:966–972.

Amweg EL, Weston DP, You J, Lydy MJ (2006) Pyrethroid insecticides and sediment toxicity in urban creeks from California and Tennessee. Environ Sci Technol 40:1700–1706.

Anderson BS, Hunt JW, Phillips BM, Fairey R, Oakden J, Puckett HM, Stephenson M, Tjeerdema RS, Long ER, Wilson CJ, Lyons M (2001) Sediment quality in Los Angeles Harbor: a triad assessment. Environ Toxicol Chem 20:359–370.

Anderson BS, de Vlaming V, Larson K, Deanovic LA, Birosik S, Smith DJ, Hunt JW, Phillips BM, Tjeerdema RS (2002) Causes of ambient toxicity in the Calleguas Creek watershed of southern California. Environ Monit Assess 78:131–151.

Anderson BS, Hunt JW, Phillips BM, Nicely PA, de Vlaming V, Connor V, Richard N, Tjeerdema RS (2003) Integrated assessment of the impacts of agricultural drainwater in the Salinas River (California, USA). Environ Pollut 124:523–532.

Anderson BS, Phillips BM, Hunt JW, Connor V, Richard N, Tjeerdema RS (2006a) Identifying primary stressors impacting macroinvertebrates in the Salinas River (California, USA): relative effects of pesticides and suspended particles. Environ Pollut 141:402–408.

Anderson BS, Phillips BM, Hunt JW, Huntley SA, Worcester K, Richard N, Tjeerdema RS (2006b) Evidence of pesticide impacts in the Santa Maria River watershed (California, U.S.). Environ Toxicol Chem 25:1160–1170.

Anderson BS, Lowe S, Hunt JW, Phillips BM, Voorhees JP, Clark SL, Tjeerdema RS (2007) Relative sensitivities of toxicity test protocols with the amphipods *Eohaustorius estuarius* and *Ampelisca abdita*. Ecotoxicol Environ Saf 69:24–31.

Anderson BS, Phillips BM, Hunt JW, Voorhees JP, Clark SL, Mekebri A, Crane D, Tjeerdema RS (in press-b) Recent advances in sediment toxicity identification evaluations emphasizing pyrethroid pesticides. In: Gan J-G, Hendley P, Spurlock F, Weston D (eds) Synthetic Pyrethroids: Occurence and Behavior in Aquatic Environments. American Chemical Society, Washington, DC, in press.

Anderson BS, Hunt JW, Phillips BM, Tjeerdema RS (2007) Navigating the TMDL Process: Sediment Toxicity. Water Environment Research Foundation, Report number 02-WSM-2 .

Ankley GT, Dierkes JR, Jensen DA, Peterson GS (1991) Piperonyl butoxide as a tool in aquatic toxicological research with organophosphate insecticides. Ecotoxicol Environ Saf 21:266–274.

Arufe MI, Arellano JM, Garcia L, Albendin G, Sarasquete C (2007) Cholinesterase activity in gilthead seabream (*Sparus aurata*) larvae: characterization and sensitivity to the organophosphate azinphosmethyl. Aquat Toxicol 84:328–336.

Attademo AM, Peltzer PM, Lajmanovich RC, Cabagna M, Fiorenza G (2007) Plasma B-esterase and glutathione S-transferase activity in the toad *Chaunus schneideri* (Amphibia, Anura) inhabiting rice agroecosystems of Argentina. Ecotoxicology 16:533–539.

Babczy ska A, Wilczek G, Migula P (2006) Effects of dimethoate on spiders from metal pollution gradient. Sci Total Environ 370:352–359.

Bacey J, Starner K, Spurlock F (2003) Preliminary Results of Study #214: Monitoring the Occurrence and Concentration of Esfenvalerate and Permethrin Pyrethroids. Department of Pesticide Regulation, Sacramento, CA.

Bacey J, Spurlock F, Starner K, Feng H, Hsu J, White J, Tran DM (2005) Residues and toxicity of esfenvalerate and permethrin in water and sediment, in tributaries of the Sacramento and San Joaquin rivers, California, USA. Bull Environ Contam Toxicol 74:864–871.

Bailey HC, Miller JL, Miller MJ, Dhaliwal BS (1995) Application of toxicity identification procedures to the echinoderm fertilization assay to identify toxicity in a municipal effluent. Environ Toxicol Chem 14:2181–2186.

Bailey HC, Digiorgio C, Kroll K, Miller JL, Hinton DE, Starrett G (1996) Development of procedures for identifying pesticide toxicity in ambient waters: carbofuran, diazinon, chlorpyrifos. Environ Toxicol Chem 15:837–845.

Bailey HC, Miller JL, Miller MJ, Wiborg LC, Deanovic LA, Shed T (1997) Joint acute toxicity of diazinon and chlorpyrifos to *Ceriodaphnia dubia*. Environ Toxicol Chem 16:2304–2308.

Bailey HC, Deanovic LA, Reyes E, Kimball T, Larson K, Cortright K, Connor V, Hinton DE (2000) Diazinon and chlorpyrifos in urban waterways in Northern California, USA. Environ Toxicol Chem 19:82–87.

Baker CMA, Manwell C, Labisky RF, Harper JA (1966) Molecular genetics of avian proteins. V. Egg, blood and tissue proteins of the ring necked pheasant *Phasianius colchius*. Comp Biochem Physiol 17:467–499.

Barata C, Solayan A, Porte C (2004) Role of B-esterases in assessing toxicity of organophosphorus (chlorpyrifos, malathion) and carbamate (carbofuran) pesticides to *Daphnia magna*. Aquat Toxicol 66:125–139.

Barata C, Damasio J, Angel López M, Kuster M, López de Alda M, Barceló D, Carmen Riva M, Raldúa D (2007) Combined use of biomarkers and *in situ* bioassays in *Daphnia magna* to monitor environmental hazards of pesticides in the field. Environ Toxicol Chem 26:370–379.

Barron MG, Charron KA, Stott WT, Duvall SE (1999) Tissue carboxylesterase activity of rainbow trout. Environ Toxicol Chem 18:2506–2511.

Bartkowiak DJ, Wilson BW (1995) Avian plasma carboxylesterase activity as a potential biomarker of organophosphate pesticide exposure. Environ Toxicol Chem 14:2149–2153.

Basack SB, Oneto ML, Fuchs JS, Wood EJ, Kesten EM (1998) Esterases of *Corbicula fluminea* as biomarkers of exposure to organophosphorus pesticides. Bull Environ Contam Toxicol 61:569–576.

Bencharit S, Morton CL, Howard-Williams EL, Danks MK, Potter PM, Redinbo MR (2002) Structural insights into CPT-11 activation by mammalian carboxylesterases. Nat Struct Biol 9:337–342.

Bencharit S, Morton CL, Hyatt JL, Kuhn P, Danks MK, Potter PM, Redinbo MR (2003a) Crystal structure of human carboxylesterase 1 complexed with the Alzheimer's drug tacrine: from binding promiscuity to selective inhibition. Chem Biol 10:341–349.

Bencharit S, Morton CL, Xue Y, Potter PM, Redinbo MR (2003b) Structural basis of heroin and cocaine metabolism by a promiscuous human drug-processing enzyme. Nat Struct Biol 10:349–356.

Bencharit S, Edwards CC, Morton CL, Howard-Williams EL, Kuhn P, Potter PM, Redinbo MR (2006) Multisite promiscuity in the processing of endogenous substrates by human carboxylesterase 1. J Mol Biol 363:201–214.

Besser JM, Ingersoll CG, Leonard EN, Mount DR (1998) Effect of zeolite on toxicity of ammonia in freshwater sediments: implications for toxicity identification evaluation procedures. Environ Toxicol Chem 17:2310–2317.

Blaise C, Ménard L (1998) A micro-algal solid-phase test to assess the toxic potential of freshwater sediments. Water Qual Res J Can 33:133–151.

Bloomquist JR, Adams PM, Soderlund DM (1986) Inhibition of gamma-aminobutyric acid-stimulated chloride flux in mouse brain vesicles by polychlorocycloalkane and pyrethroid insecticides. Neurotoxicology 7:11–20.

Bodor N, Buchwald P (2000) Soft drug design: general principles and recent applications. Med Res Rev 20:58–101.

Bodor N, Buchwald P (2003) Retrometabolism-based drug design and targeting. In: Abraham D (ed) Burger's Medicinal Chemistry and Drug Discovery, 6th Ed. Vol 2, Wiley, New York, pp 534–596.

Bodor N, Buchwald P (2004) Designing safer (soft) drugs by avoiding the formation of toxic and oxidative metabolites. Mol Biotechnol 26:123–132.

Bonacci S, Browne MA, Dissanayake A, Hagger JA, Corsi I, Focardi S, Galloway TS (2004) Esterase activities in the bivalve mollusc *Adamussium colbecki* as a biomarker for pollution monitoring in the Antarctic marine environment. *Mar Pollut Bull* 49:445–455.

Bond J-A, Bradley BP (1997) Resistance to malathion in heat-shocked *Daphnia magna*. Environ Toxicol Chem 16:705–712.

Bondarenko S, Putt A, Kavanaugh S, Poletika N, Gan J (2006) Time dependence of phase distribution of pyrethroid insecticides in sediment. Environ Toxicol Chem 25:3148–3154.

Bornscheuer UT (2002) Microbial carboxyl esterases: classification, properties and application in biocatalysis. FEMS Microbiol Rev 26:73–81.

Bradberry SM, Cage SA, Proudfoot AT, Vale JA (2005) Poisoning due to pyrethroids. Toxicol Rev 24:93–106.

Bradbury SP, Coats JR (1989a) Toxicokinetics and toxicodynamics of pyrethroid insecticides in fish. Environ Toxicol Chem 8:373–380.

Bradbury SP, Coats JR (1989b) Comparative toxicology of the pyrethroid insecticides. Rev Environ Contam Toxicol 108:133–177.

Brodbeck U, Schweikert K, Gentinetta R, Rottenberg M (1979) Fluorinated aldehydes and ketones acting as quasi-substrate inhibitors of acetylcholinesterase. Biochim Biophys Acta 567:357–369.

Brooks GT (1986) Insecticide metabolism and selective toxicity. Xenobiotica 16:989–1002.

Burgess RM, Charles JB, Kuhn A, Ho KT, Patton LE, McGovern DG (1997) Development of a cation-exchange methodology for marine toxicity identification evaluation applications. Environ Toxicol Chem 16:1203–1211.

Burgess RM, Cantwell MG, Pelletier MC, Ho KT, Serbst JR, Cook H, Kuhn A (2000) Development of a toxicity identification evaluation procedure for characterizing metal toxicity in marine sediments. Environ Toxicol Chem 19:982–991.

Burgess RM, Pelletier MC, Ho KT, Serbst JR, Ryba SA, Kuhn A, Perron MM, Raczelowski P, Cantwell MG (2003) Removal of ammonia toxicity in marine sediment TIEs: a comparison of *Ulva lactuca*, zeolite and aeration methods. Mar Pollut Bull 46:607–618.

Burgess RM, Perron MM, Cantwell MG, Ho KT, Serbst JR, Pelletier MC (2004) Use of zeolite for removing ammonia and ammonia-caused toxicity in marine toxicity identification evaluations. Arch Environ Contam Toxicol 47:440–447.

Butte W, Kemper K (1999) A spectrophotometric assay for pyrethroid-cleaving enzymes in human serum. Toxicol Lett 107:4953.

Byrne FJ, Gorman KJ, Cahill M, Denholm I, Devonshire AL (2000) The role of B-type esterases in conferring insecticide resistance in the tobacco whitefly, *Bemisia tabaci* (Genn). Pestic Manag Sci 56:867–874.

Cahill M, Byrne FJ, German K, Denholm I, Devonshire AL (1995) Pyrethroid and organophosphate resistance in the Tobacco Whitefly *Bemisia tabaci* (Homoptera, Aleyrodidae). B Entomol Res 85:181–187.

Casida JE (1970) Mixed-function oxidase involvement in the biochemistry of insecticide synergists. J Agric Food Chem 18:753–772.

Casida JE (1973) Pyrethrum, the Natural Insecticide. Academic Press, New York.

Casida JE (1980) Pyrethrum flowers and pyrethroid insecticides. Environ Health Perspect 34:189–202.

Casida JE, Quistad GB (1995) Metabolism and synergism of pyrethrins. In: Casida JE, Quistad GB (eds) Pyrethrum Flowers: Production, Chemistry, Toxicology, and Uses, 1st Ed. Oxford University Press, New York, pp 258–276.

Casida JE, Quistad GB (1998) Golden age of insecticide research: past, present, or future? Annu Rev Entomol 43:1–16.

Casida JE, Quistad GB (2004) Organophosphate toxicology: safety aspects of nonacetylcholinesterase secondary targets. Chem Res Toxicol 17:983–998.

Casida JE, Quistad GB (2005) Serine hydrolase targets of organophosphorus toxicants. Chem Biol Interact 157–158:277–283.

Casida JE, Gammon DW, Glickman AH, Lawrence LJ (1983) Mechanisms of selective action of pyrethroid insecticides. Annu Rev Pharmacol Toxicol 23:413–438.

Castellanos LC, Sanchez-Hernandez JC (2007) Earthworm biomarkers of pesticide contamination: current status and perspectives. J Pestic Sci 32:360–371.

Chambers JE (1976) The relationship of esterases to organophosphorus insecticide tolerance in mosquitofish. Pestic Biochem Physiol 6:517–522.

Chevre N, Gagne F, Blaise C (2003) Development of a biomarker-based index for assessing the ecotoxic potential of aquatic sites. Biomarkers 8:287–298.

Chiang SW, Sun CN (1996) Purification and characterization of carboxylesterases of rice green leafhopper *Nephotettix cincticeps* Uhler. Pest Biochem Physiol 54:181–189.

Choi J, Hodgson E, Rose RL (2004) Inhibition of *trans*-permethrin hydrolysis in human liver fractions by chloropyrifos oxon and carbaryl. Drug Metab Drug Interact 20:233–246.

Coats JR (1990) Mechanisms of toxic action and structure-activity relationships for organochlorine and synthetic pyrethroid insecticides. Environ Health Perspect 87:255–262.

Coats JR, Symonik DM, Bradbury SP, Dyer SD, Timson LK, Atchison GJ (1989) Toxicology of synthetic pyrethroids in aquatic organisms: an overview. Environ Toxicol Chem 8:671–679.

Coleman M, Hemingway J (2007) Insecticide resistance monitoring and evaluation in disease transmitting mosquitoes. J Pestic Sci 32:69–76.
Cordi B, Fossi C, Depledge MH (1997) Temporal biomarker responses in wild passerine birds exposed to pesticide spray drift. Environ Toxicol Chem 16:2118–2124.
Corsi I, Bonacci S, Santovito G, Chiantore M, Castagnolo L, Focardi S (2004) Cholinesterase activities in the Antarctic scallop *Adamussium colbecki*: tissue expression and effect of $ZnCl_2$ exposure. Mar Environ Res 58:401–406.
Cui F, Weill M, Berthomieu A, Raymond M, Qiao CL (2007) Characterization of novel esterases in insecticide-resistant mosquitoes. Insect Biochem Mol Biol 37:1131–1137.
Curtis CF, Mnzava AE (2000) Comparison of house spraying and insecticide-treated nets for malaria control. Bull W H O 78:1389–1400.
Cygler M, Schrag JD, Sussman JL, Harel M, Silman I, Gentry MK, Doctor BP (1993) Relationship between sequence conservation and three-dimensional structure in a large family of esterases, lipases, and related proteins. Protein Sci 2:366–382.
Davies JH (1985) The pyrethroids: an historical introduction. In: Leahey JP (ed) The Pyrethroid Insecticides. Taylor & Francis, Philadelphia, pp 1–41.
Day KE (1989) Acute, chronic and sublethal effects of synthetic pyrethroids on freshwater zooplankton. Environ Toxicol Chem 8:411–416.
Denton D (2001) Integrated Toxicological and Hydrological Assessments of Diazinon and Esfenvalerate. PhD dissertation. University of California Davis, Davis.
Denton DL, Wheelock CE, Murray S, Deanovic LA, Hammock BD, Hinton DE (2003) Joint acute toxicity of esfenvalerate and diazinon to fathead minnow (*Pimephales promelas*) larvae. Environ Toxicol Chem 22:336–341.
Dettbarn WD, Yang ZP, Milatovic D (1999) Different role of carboxylesterases in toxicity and tolerance to paraoxon and DFP. Chem Biol Interact 119–120:445–454.
de Vlaming V, Connor V, DeGiorgio C, Bailey HC, Deanovic LA, Hinton DE (2000) Application of whole effluent toxicity test procedures to ambient water quality assessment. Environ Toxicol Chem 19:42–62.
Devonshire AL, Heidari R, Huang HZ, Hammock BD, Russell RJ, Oakeshott JG (2007) Hydrolysis of individual isomers of fluorogenic pyrethroid analogs by mutant carboxylesterases from *Lucilia cuprina*. Insect Biochem Mol Biol 37:891–902.
Elliott M (1976) Properties and applications of pyrethroids. Environ Health Perspect 14:1–2.
Elliott M, Janes NF, Jeffs KA, Needham PH, Sawicki RM (1965) New pyrethrin-like esters with high insecticidal activity. Nature (Lond) 207:938–940.
Elliott M, Farnham AW, Janes NF, Needham PH, Pearson BC (1967) 5-Benzyl-3-furylmethyl chrysanthemate: a new potent insecticide. Nature (Lond) 213:493–494.
Elliott M, Farnham AW, Janes NF, Needham PH, Pulman DA (1973a) Potent pyrethroid insecticides from modified cyclopropane acids. Nature (Lond) 244:456–457.
Elliott M, Farnham AW, Janes NF, Needham PH, Pulman DA, Stevenson JH (1973b) A photostable pyrethroid. Nature (Lond) 246:169–170.
Elliott M, Farnham AW, Janes NF, Needham PH, Pulman DA (1974) Synthetic insecticide with a new order of activity. Nature (Lond) 248:710–711.
Elzen GW, Leonard BR, Graves JB, Burris E, Micinski S (1992) Resistance to pyrethroid, carbamate, and organophosphate insecticides in field populations of Tobacco Budworm (Lepidoptera, Noctuidae) in 1990. J Econ Entomol 85:2064–2072.
Enan E, Matsumura F (1993) Activation of phosphoinositide/protein kinase C pathway in rat brain tissue by pyrethroids. Biochem Pharmacol 45:703–710.
Enayati AA, Ranson H, Hemingway J (2005) Insect glutathione transferases and insecticide resistance. Insect Mol Biol 14:3–8.
Epstein L, Bassein S, Zalom FG (2000) Almond and stone fruit growers reduce OP, increase pyrethroid use in dormant sprays. Calif Agric 54:14–19.
Escartin E, Porte C (1997) The use of cholinesterase and carboxylesterase activities from *Mytilus galloprovincialis* in pollution monitoring. Environ Toxicol Chem 16:2090–2095.

Ferrari A, Venturino A, Pechén de D'Angelo AM (2007) Effects of carbaryl and azinphos-methyl on juvenile rainbow trout (*Oncorhynchus mykiss*) detoxifying enzymes. Pestic Biochem Physiol 88:134–142.

Fleming CD, Bencharit S, Edwards CC, Hyatt JL, Tsurkan L, Bai F, Fraga C, Morton CL, Howard-Williams EL, Potter PM, Redinbo MR (2005) Structural insights into drug processing by human carboxylesterase. 1: Tamoxifen, mevastatin, and inhibition by benzil. J Mol Biol 352:165–177.

Fleming CD, Edwards CC, Kirby SD, Maxwell DM, Potter PM, Cerasoli DM, Redinbo MR (2007) Crystal structures of human carboxylesterase 1 in covalent complexes with the chemical warfare agents Soman and Tabun. Biochemistry 46:5063–5071.

Forshaw PJ, Lister T, Ray DE (1993) Inhibition of a neuronal voltage-dependent chloride channel by the type II pyrethroid, deltamethrin. Neuropharmacology 32:105–111.

Fossi MC, Leonzio C, Massi A, Lari L, Casini S (1992) Serum esterase inhibition in birds: a non-destructive biomarker to assess organophosphorus and carbamate contamination. Arch Environ Contam Toxicol 23:99–104.

Fossi MC, Massi A, Leonzio C (1994) Blood esterase inhibition in birds as an index of organophosphorus contamination: field and laboratory studies. Ecotoxicology 3:11–20.

Fossi MC, Sanchez-Hernandez JC, Diaz-Diaz R, Lari L, Garcia-Hernandez JE, Gaggi C (1995) The lizard *Gallotia galloti* as a bioindicator of organophosphorus contamination in the Canary Islands. Environ Pollut 87:289–294.

Fossi MC, Lari L, Casini S (1996) Interspecies variation of "B" esterases in birds: the influence of size and feeding habits. Arch Environ Contam Toxicol 31:525532.

Fourcy D, Jumel A, Heydorff M, Lagadic L (2002) Esterases as biomarkers in *Nereis* (*Hediste*) *diversicolor* exposed to temephos and *Bacillus thuringiensis* var. *israelensis* used for mosquito control in coastal wetlands of Morbihan (Brittany, France). Mar Environ Res 54:755–759.

Fournier D, Bride JM, Poirie M, Berge JB, Plapp FW Jr (1992) Insect glutathione S-transferases. Biochemical characteristics of the major forms from houseflies susceptible and resistant to insecticides. J Biol Chem 267:1840–1845.

Fujitani Y (1909) Chemistry and pharmacology of insect powder. Arch Exp Pathol Pharmacol 61:47–75.

Fukuto TR (1990) Mechanism of action of organophosphorus and carbamate insecticides. Environ Health Perspect 87:245–254.

Fulton MH, Key PB (2001) Acetylcholinesterase inhibition in estuarine fish and invertebrates as an indicator of organophosphorus insecticide exposure and effects. Environ Toxicol Chem 20:37–45.

Galloway TS (2006) Biomarkers in environmental and human health risk assessment. Mar Pollut Bull 53:606–613.

Galloway TS, Millward N, Browne MA, Depledge MH (2002) Rapid assessment of organophosphorous/carbamate exposure in the bivalve mollusc *Mytilus edulis* using combined esterase activities as biomarkers. Aquat Toxicol 61:169–180.

Galloway TS, Brown RJ, Browne MA, Dissanayake A, Lowe D, Jones MB, Depledge MH (2004a) Ecosystem management bioindicators: the ECOMAN project: a multi-biomarker approach to ecosystem management. Mar Environ Res 58:233–237.

Galloway TS, Brown RJ, Browne MA, Dissanayake A, Lowe D, Jones MB, Depledge MH (2004b) A multibiomarker approach to environmental assessment. Environ Sci Technol 38:1723–1731.

Galloway TS, Brown RJ, Browne MA, Dissanayake A, Lowe D, Depledge MH, Jones MB (2006) The ECOMAN project: A novel approach to defining sustainable ecosystem function. Mar Pollut Bull 53:186–194.

Gan J, Lee SJ, Liu WP, Haver DL, Kabashima JN (2005) Distribution and persisitence of pyrethroids in runoff sediments. J Environ Qual 34:836–841.

Gaughan LC, Engel JL, Casida JE (1980) Pestcide interactions: effects of organophosphorus pesticides on the metabolism, toxicity, and persistence of selected pyrethroid insecticides. Pestic Biochem Physiol 14:81–85.

Gershater M, Sharples K, Edwards R (2006) Carboxylesterase activities toward pesticide esters in crops and weeds. Phytochemistry 67:2561–2567.

Glade MJ (1998) The Food Quality Protection Act of 1996. Nutrition 14:65–66.

Glickman AH, Casida JE (1982) Species and structural variations affecting pyrethroid neurotoxicity. Neurobehav Toxicol Teratol 4:793–799.

Glickman AH, Lech JJ (1981) Hydrolysis of permethrin, a pyrethroid insecticide, by rainbow trout and mouse tissues *in vitro*: a comparative study. Toxicol Appl Pharmacol 60:186–192.

Glickman AH, Lech JJ (1982) Differential toxicity of trans-permethrin in rainbow trout and mice. II. Role of target organ sensitivity. Toxicol Appl Pharmacol 66:162–171.

Glickman AH, Shono T, Casida JE, Lech JJ (1979) *In vitro* metabolism of permethrin isomers by carp and rainbow trout liver microsomes. J Agric Food Chem 27:1038–1041.

Glickman AH, Weitman SD, Lech JJ (1982) Differential toxicity of *trans*-permethrin in rainbow trout and mice. I. Role of biotransformation. Toxicol Appl Pharmacol 66:153–161.

Greenwood BM, Bojang K, Whitty CJ, Targett GA (2005) Malaria. Lancet 365:1487–1498.

Gupta RC, Dettbarn WD (1993) Role of carboxylesterases in the prevention and potentiation of N-methylcarbamate toxicity. Chem-Biol Interact 87:295–303.

Gupta RC, Kadel WL (1990) Toxic interaction of tetraisopropylpyrophosphoramide and propoxur: some insights into the mechanisms. Arch Environ Contam Toxicol 19:917–920.

Hamers T, Molin KR, Koeman JH, Murk AJ (2000) A small-volume bioassay for quantification of the esterase inhibiting potency of mixtures of organophosphate and carbamate insecticides in rainwater: development and optimization. Toxicol Sci 58:60–67.

Hamers T, van den Brink PJ, Mos L, van der Linden SC, Legler J, Koeman JH, Murk AJ (2003) Estrogenic and esterase-inhibiting potency in rainwater in relation to pesticide concentrations, sampling season and location. Environ Pollut 123:47–65.

Hamm JT, Wilson BW, Hinton DE (2001) Increasing uptake and bioactivation with development positively modulate diazinon toxicity in early life stage medaka (*Oryzias latipes*). Toxicol Sci 61:304–313.

Hannam ML, Hagger JA, Jones MB, Galloway TS (in press) Characterisation of esterases as potential biomarkers of pesticide exposure in the lugworm *Arenicola marina* (Annelida: Polychaeta). Environ Pollut in press.

Harold JA, Ottea JA (2000) Characterization of esterases associated with profenofos resistance in the tobacco budworm, *Heliothis virescens* (F.). Arch Insect Biochem Physiol 45:47–59.

He F, Wang S, Liu L, Chen S, Zhang Z, Sun J (1989) Clinical manifestations and diagnosis of acute pyrethroid poisoning. Arch Toxicol 63:54–58.

Heikinheimo P, Goldman A, Jeffries C, Ollis DL (1999) Of barn owls and bankers: a lush variety of alpha/beta hydrolases. Structure Fold Descr 7:R141–146.

Hemingway J, Karunaratne SH (1998) Mosquito carboxylesterases: a review of the molecular biology and biochemistry of a major insecticide resistance mechanism. Med Vet Entomol 12:1–12.

Hicks LD, Hyatt JL, Moak T, Edwards CC, Tsurkan L, Wierdl M, Ferreira AM, Wadkins RM, Potter PM (2007) Analysis of the inhibition of mammalian carboxylesterases by novel fluorobenzoins and fluorobenzils. Bioorg Med Chem 15:3801–3817.

Ho KT, Kuhn A, Pelletier MC, McGee F, Burgess RM, Serbst JR (2000) Sediment toxicity assessment: comparison of standard and new testing designs. Arch Environ Contam Toxicol 39:462–468.

Ho KT, Kuhn A, Pelletier MC, Serbst JR, Cook H, Cantwell MG, Ryba SA, Perron MM, Lebo J, Huckins J, Petty J (2004) Use of powdered coconut charcoal as a toxicity identification and evaluation manipulation for organic toxicants in marine sediments. Environ Toxicol Chem 23:2124–2131.

Hodgson E (1982) Production of pesticide metabolites by oxidative reactions. J Toxicol Clin Toxicol 19:609–621.

Hosokawa M, Endo T, Fujisawa M, Hara S, Iwata N, Sato Y, Satoh T (1995) Interindividual variation in carboxylesterase levels in human liver microsomes. Drug Metab Dispos 23:1022–1027.

Hosokawa M, Furihata T, Yaginuma Y, Yamamoto N, Koyano N, Fujii A, Nagahara Y, Satoh T, Chiba K (2007) Genomic structure and transcriptional regulation of the rat, mouse, and human carboxylesterase genes. Drug Metab Rev 39:1–15.

Hotelier T, Renault L, Cousin X, Negre V, Marchot P, Chatonnet A (2004) ESTHER, the database of the alpha/beta-hydrolase fold superfamily of proteins. Nucleic Acids Res 32(database issue):D145–D147.

Huang H, Ottea JA (2004) Development of pyrethroid substrates for esterases associated with pyrethroid resistance in the tobacco budworm, *Heliothis virescens* (F.). J Agric Food Chem 52:6539–6545.

Huang H, Fleming CD, Nishi K, Redinbo MR, Hammock BD (2005) Stereoselective hydrolysis of pyrethroid-like fluorescent substrates by human and other mammalian liver carboxylesterases. Chem Res Toxicol 18:1371–1377.

Huang H, Nishi K, Gee SJ, Hammock BD (2006) Evaluation of chiral alpha-cyanoesters as general fluorescent substrates for screening enantioselective esterases. J Agric Food Chem 54:694–699.

Huang TL, Obih PO, Jaiswal R, Hartley WR, Thiyagarajah A (1997) Evaluation of liver and brain esterases in the spotted gar fish (*Lepisosteus oculatus*) as biomarkers of effect in the lower Mississippi River Basin. Bull Environ Contam Toxicol 58:688–695.

Hunt JW, Anderson BS, Phillips BM, Tjeerdema RS, Taberski KM, Wilson CJ, Puckett HM, Stephenson M, Fairey R, Oakden J (2001) A large-scale categorization of sites in San Francisco Bay, USA, based on the sediment quality triad, toxicity identification evaluations, and gradient studies. Environ Toxicol Chem 20:1252–1265.

Hunt JW, Anderson BS, Phillips BM, Nicely PN, Tjeerdema RS, Puckett HM, Stephenson M, Worcester K, De Vlaming V (2003) Ambient toxicity due to chlorpyrifos and diazinon in a central California coastal watershed. Environ Monit Assess 82:83–112.

Hyatt JL, Stacy V, Wadkins RM, Yoon KJ, Wierdl M, Edwards CC, Zeller M, Hunter AD, Danks MK, Crundwell G, Potter PM (2005) Inhibition of carboxylesterases by benzil (diphenylethane-1,2-dione) and heterocyclic analogues is dependent upon the aromaticity of the ring and the flexibility of the dione moiety. J Med Chem 48:5543–5550.

Hyatt JL, Moak T, Hatfield MJ, Tsurkan L, Edwards CC, Wierdl M, Danks MK, Wadkins RM, Potter PM (2007) Selective inhibition of carboxylesterases by isatins, indole-2,3-diones. J Med Chem 50:1876–1885.

Hyne RV, Maher WA (2003) Invertebrate biomarkers: links to toxicosis that predict population decline. Ecotoxicol Environ Saf 54:366–374.

Ileperuma NR, Marshall SD, Squire CJ, Baker HM, Oakeshott JG, Russell RJ, Plummer KM, Newcomb RD, Baker EN (2007) High-resolution crystal structure of plant carboxylesterase AeCXE1, from *Actinidia eriantha*, and its complex with a high-affinity inhibitor paraoxon. Biochemistry 46:1851–1859.

Imai T (2006) Human carboxylesterase isozymes: catalytic properties and rational drug design. Drug Metab Pharmacokinet 21:173–185.

Immaraju JA, Morse JG, Gaston LK (1990) Mechanisms of organophosphate, pyrethroid, and DDT resistance in Citrus Thrips (Thysanoptera, Thripidae). J Econ Entomol 83:1723–1732.

James MO (1986) Overview of *in vitro* metabolism of drugs by aquatic species. Vet Hum Toxicol 28(suppl 1):2–8.

Junge W, Krisch K (1975) The carboxylesterases/amidases of mammalian liver and their possible significance. Crit Rev Food Sci 371:434.

Kakko I, Toimela T, Tähti H (2000) Piperonyl butoxide potentiates the synaptosome ATPase inhibiting effect of pyrethrin. Chemosphere 40:301–305.

Kao L, Motoyama N, Dauterman W (1985) Multiple forms of esterases in mouse, rat, and rabbit liver, and their role in hydrolysis of organophosphate and pyrethroid insecticides. Pestic Biochem Physiol 23:66–73.

Karpouzas DG, Singh BK (2006) Microbial degradation of organophosphorus xenobiotics: metabolic pathways and molecular basis. Adv Microb Physiol 51:119–1185

Katsuda Y (1999) Development of and future prospects for pyrethroid chemistry. Pestic Sci 55:775–782.

Keizer J, D'Agostino G, Nagel R, Volpe T, Gnemi P, Vittozzi L (1995) Enzymological differences of AChE and diazinon hepatic metabolism: correlation of *in vitro* data with the selective toxicity of diazinon to fish species. Sci Total Environ 171:213–220.

Kelley K, Starner K (2004) Preliminary Results for Study 219: Monitoring Surface Waters and Sediments of the Salinas and San Joaquin River Basins for Synthetic Pyrethroid Pesticides. Department of Pesticide Regulation, Sacramento, CA.

Kingsbury N, Masters CJ (1972) Heterogeneity, molecular weight interrelationships and developmental genetics of the esterase isoenzymes of the rainbow trout. Biochim Biophys Acta 258:455–465.

Kosian PA, West CW, Pasha MS, Cox JS, Mount DR, Huggett RJ, Ankley GT (1999) Use of nonpolar resin for reduction of fluoranthene bioavailability in sediment. Environ Toxicol Chem 18:201–206.

Kulkarni AP, Hodgson E (1984) The metabolism of insecticides: the role of monooxygenase enzymes. Annu Rev Pharmacol Toxicol 24:19–42.

Kuster E (2005) Cholin- and carboxyl-esterase activities in developing zebrafish embryos (*Danio rerio*) and their potential use for insecticide hazard assessment. Aquat Toxicol 75:76–85.

Kuster E, Altenburger R (2006) Comparison of cholin- and carboxyl-esterase enzyme inhibition and visible effects in the zebra fish embryo bioassay under short-term paraoxon-methyl exposure. Biomarkers 11:341–354.

LaForge FB, Barthel WF (1945) Constituents of pyrethrum flowers. XVIII. The structure and isomerism of pyrethrolone and cinerolone. J Org Chem 10:114–120.

Lari L, Massi A, Fossi MC, Casini S, Leonzio C, Focardi S (1994) Evaluation of toxic effects of the organophosphorus insecticide azinphos-methyl in experimentally and naturally exposed birds. Arch Environ Contam Toxicol 26:234–239.

Laskowski DA (2002) Physical and chemical properties of pyrethroids. Rev Environ Contam Toxicol 174:49–170.

Lebo JA, Huckins JN, Petty JD, Ho KT (1999) Removal of organic contaminant toxicity from sediments: early work toward development of a toxicity identification evaluation (TIE) method. Chemosphere 39:389–406.

Lebo JA, Huckins JN, Petty JD, Cranor WL, Ho KT (2003) Comparisons of coarse and fine versions of two carbons for reducing the bioavailabilities of sediment-bound hydrophobic organic contaminants. Chemosphere 50:1309–1317.

Lee H-J, Shan G, Watanabe T, Stoutamire DW, Gee SJ, Hammock BD (2002) Enzyme-linked immunosorbent assay for the pyrethroid deltamethrin. J Agric Food Chem 50:5526–5532.

Lee H-J, Shan G, Ahn K, Park E-K, Watanabe T, Gee SJ, Hammock BD (2004) Development of an enzyme-linked immunosorbent assay for the pyrethroid cypermethrin. J Agric Food Chem 52:1039–1043.

Lee S, Gan J, Kabashima J (2002) Recovery of synthetic pyrethroids in water samples during storage and extraction. J Agric Food Chem 50:7194–7198.

Leng G, Lewalter J, Rohrig B, Idel H (1999) The influence of individual susceptibility in pyrethroid exposure. Toxicol Lett 107:123–130.

Li SN, Fan DF (1996) Correlation between biochemical parameters and susceptibility of freshwater fish to malathion. J Toxicol Environ Health 48:413–418.

Li SN, Fan DF (1997) Activity of esterases from different tissues of freshwater fish and responses of their isoenzymes to inhibitors. J Toxicol Environ Health 51:149–157.

Liu W, Gan JJ, Lee S, Kabashima JN (2004) Phase distribution of synthetic pyrethroids in runoff and stream water. Environ Toxicol Chem 23:7–11.

Lydy MJ, Beldin J, Wheelock CE, Hammock BD, Denton DL (2004) Challenges in regulating pesticide mixtures. Ecology and Society 9:1. [online] http://www.ecologyandsociety.org/vol9/iss6/art1.

Mak SK, Shan G, Lee H-J, Watanabe T, Stoutamire DW, Gee SJ, Hammock BD (2005) Development of a class selective immunoassay for the type II pyrethroid insecticides. Anal Chim Acta 534:109–120.

Martin T, Ochou OG, Vaissayre M, Fournier D (2003) Organophosphorus insecticides synergize pyrethroids in the resistant strain of cotton bollworm, *Helicoverpa armigera* (Hubner) (Lepidoptera: Noctuidae) from West Africa. J Econ Entomol 96:468–474.

Maund SJ, Hamer MJ, Lane MCG, Farrelly E, Rapley JH, Goggin UM, Gentle WE (2002) Partitioning, bioavailability, and toxicity of the pyrethroid insecticide cypermethrin in sediments. Environ Toxicol Chem 21:9–15.

Maxwell DM (1992a) Detoxification of organophosphorus compounds by carboxylesterase. In: Chambers JE, Levi PE (eds) Organophosphates: Chemistry, Fate, and Effects. Academic Press, San Diego, pp 183–199.

Maxwell DM (1992b) The specificity of carboxylesterase protection against the toxicity of organophosphate compounds. Toxicol Appl Pharmacol 114:306–312.

Maxwell DM, Lieske CN, Brecht KM (1994) Oxime-induced reactivation of carboxylesterase inhibited by organophosphorus compounds. Chem Res Toxicol 7:428–433.

Maxwell DM, Brecht KM (2001) Carboxylesterase: specificity and spontaneous reactivation of an endogenous scavenger for organophosphorus compounds. J Appl Toxicol 21(suppl 1):S103–S107.

Mazzarri MB, Georghiou GP (1995) Characterization of resistance to organophosphate, carbamate, and pyrethroid insecticides in field populations of *Aedes aegypti* from Venezuela. J Am Mosquito Control Assoc 11:315–322.

McAbee RD, Kang K, Stanich MA, Christiansen J, Wheelock CE, Inman A, Hammock BD, Cornel AJ (2004) Pyrethroid tolerance in *Culex pipiens pipiens* var. *molestus* from Marin County, California. Pestic Manag Sci 60:359–368.

Mileson BE, Chambers JE, Chen WL, Dettbarn W, Ehrich M, Eldefrawi AT, Gaylor DW, Hamernik K, Hodgson E, Karczmar AG, Padilla S, Pope CN, Richardson RJ, Saunders DR, Sheets LP, Sultatos LG, Wallace KB (1998) Common mechanism of toxicity: a case study of organophosphorus pesticides. Toxicol Sci 41:8–20.

Moore A, Waring CP (2001) The effects of a synthetic pyrethroid pesticide on some aspects of reproductions in Atlantic salmon (*Salmo salar* L.). Aquat Toxicol 52:1–12.

Morisseau C, Hammock BD (2005) Epoxide hydrolases: mechanisms, inhibitor designs, and biological roles. Annu Rev Pharmacol Toxicol 45:311–333.

MPSL (2006a) Toxicity Identification results: Region 5: DPCCR (541STC533) and WWNCR (541STC029), Surface Water Ambient Monitoring Program. Prepared by the University of California, Davis, for the Central Valley Regional Water Quality Control Board, Sacramento, CA.

MPSL (2006b) Toxicity Identification Evaluation Results: Region 5: Grayson Drain—541STC030, Surface Water Ambient Monitoring Program. Prepared by the University of California, Davis, for the Central Valley Regional Water Quality Control Board, Sacramento, CA.

Myers M, Richmond RC, Oakeshott JG (1988) On the origins of esterases. Mol Biol Evol 5:113–119.

Narahashi T (1996) Neuronal ion channels as the target sites of insecticides. Pharmacol Toxicol 79:1–14.

Newcomb RD, Campbell PM, Ollis DL, Cheah E, Russell RJ, Oakeshott JG (1997) A single amino acid substitution converts a carboxylesterase to an organophosphorus hydrolase and confers insecticide resistance on a blowfly. Proc Natl Acad Sci U S A 94:7464–7468.

Newman JW, Morisseau C, Hammock BD (2005) Epoxide hydrolases: their roles and interactions with lipid metabolism. Prog Lipid Res 44:1–51.

Oakeshott JG, Claudianos C, Russell RJ, Robin GC (1999) Carboxyl/cholinesterases: a case study of the evolution of a successful multigene family. BioEssays 21:1031–1042.

Oakeshott JG, Devonshire AL, Claudianos C, Sutherland TD, Horne I, Campbell PM, Ollis DL, Russell RJ (2005) Comparing the organophosphorus and carbamate insecticide resistance mutations in cholin- and carboxyl-esterases. Chem Biol Interact 157–158:269–275.

O'Connor TP (2002) National distribution of chemical concentrations in mussels and oysters in the USA. Mar Environ Res 53:117–143.

Ohno N, Fujimoto K, Okuno Y, Mizutani T, Hirano M, Itaya N, Honda T, Yoshioka H (1976) 2-Arylalkanoates, a new group of synthetic pyrethroid esters not containing cyclopropanecarboxylates. Pestic Sci 7:241–246.

Ollis DL, Cheah E, Cygler M, Dijkstra B, Frolow F, Franken SM, Harel M, Remington SJ, Silman I, Schrag J, Sussman JL, Verschueren KHG, Goldman A (1992) The α/β hydrolase fold. Protein Eng 5:197–211.

O'Malley M (1997) Clinical evaluation of pesticide exposure and poisonings. Lancet 349:1161–1166.
O'Neill AJ, Galloway TS, Browne MA, Dissanayake A, Depledge MH (2004) Evaluation of toxicity in tributaries of the Mersey estuary using the isopod *Asellus aquaticus* (L.). Mar Environ Res 58:327–331.
Oros D, Werner I (2005) Pyrethroid Insecticides: An Analysis of Use Patterns, Distributions, Potential Toxicity and Fate in the Sacramento-San Joaquin Delta and Central Valley. White Paper for the Interagency Ecological Program. SFEI Contribution 415. San Francisco Estuary Institute, Oakland, CA.
Ozretic B, Krajnovic-Ozretic M (1992) Esterase heterogeneity in mussel *Mytilus galloprovincialis*: effects of organophosphate and carbamate pesticides *in vitro*. Comp Biochem Physiol C 103:221–225.
Pap L (2003) Pyrethroids. In: Plimmer JR, Gammon DW, Ragsdale NN (eds) Encyclopedia of Agrochemicals. Wiley, New York. http://www.knovel.com/knovel2/Toc.jsp?BookID=964&VerticalID=960.
Parker ML, Goldstein MI (2000) Differential toxicities of organophosphate and carbamate insecticides in the nestling European starling (*Sturnus vulgaris*). Arch Environ Contam Toxicol 39:233–242.
Pathiratne A, George SG (1998) Toxicity of malathion to Nile tilapia, *Oreochromis niloticus* and modulation by other environmental contaminants. Aquat Toxicol 43:261–271.
Phillips BM, Anderson BS, Hunt JW, Nicely PA, Kosaka RA, Tjeerdema RS, de Vlaming V, Richard N (2004) *In situ* water and sediment toxicity in an agricultural watershed. Environ Toxicol Chem 23:435–442.
Phillips BM, Anderson BA, Hunt JW, Huntley SA, Tjeerdema RS, Richard N, Worcester K (2006) Solid-phase sediment Toxicity Identification Evaluation in an agricultural stream. Environ Toxicol Chem 25:1671–1676.
Phillips BM, Anderson BS, Hunt JW, Tjeerdema RS, Carpio-Obeso M, Connor V (2007) Causes of water column toxicity to *Hyalella azteca* in the New River, California USA. Environ Toxicol Chem 26:1074–1079.
Pindel EV, Kedishvili NY, Abraham TL, Brzezinski MR, Zhang J, Dean RA, Borson WF (1997) Purification and cloning of a broad substrate specificity human liver carboxylesterase that catalyzes the hydrolysis of cocaine and heroin. J Biol Chem 272:14769–14775.
Potter PM, Wadkins RM (2006) Carboxylesterases: detoxifying enzymes and targets for drug therapy. Curr Med Chem 13:1045–1054.
Potter PM, Pawlik CA, Morton CL, Naeve CW, Danks MK (1998) Isolation and partial characterization of a cDNA encoding a rabbit liver carboxylesterase that activates the prodrug irinotecan (CPT-11). Cancer Res 58:2646–2651.
Quinn DM (1987) Acetylcholinesterase: enzyme structure, reaction dynamics, and virtual transition states. Chem Rev 87:955–979.
Quinn DM (1997) Esterases of the α/β hydrolase fold family. In: Guengerich F (ed) Biotransformation. Elsevier, Oxford, pp 243–264.
Quinn DM (1999) Ester hydrolysis. In: Poulter C (ed) Enzymes, Enzyme Mechanisms, Proteins, and Aspects of NO Chemistry. Elsevier, Oxford, pp 101–137.
Ray DE, Forshaw PJ (2000) Pyrethroid insecticides: poisoning syndromes, synergies, and therapy. J Toxicol Clin Toxicol 38:95–101.
Redinbo MR, Potter PM (2005) Mammalian carboxylesterases: from drug targets to protein therapeutics. Drug Discov Today 10:313–325.
Redinbo MR, Bencharit S, Potter PM (2003) Human carboxylesterase. 1: From drug metabolism to drug discovery. Biochem Soc Trans 31:620–624.
Regel RH, Ferris JM, Ganf GG, Brookes JD (2002) Algal esterase activity as a biomeasure of environmental degradation in a freshwater creek. Aquat Toxicol 59:209–223.
Rickwood CJ, Galloway TS (2004) Acetylcholinesterase inhibition as a biomarker of adverse effect. A study of *Mytilus edulis* exposed to the priority pollutant chlorfenvinphos. Aquat Toxicol 67:45–56.
Riddles PW, Schnitzerling HJ, Davey PA (1983) Application of trans and cis isomers of *p*-nitrophenyl-(1*R*, *S*)-3-(2,2-dichlorovinyl)-2,2-dimethylcyclopropanecarboxylate to the assay of pyrethroid hydrolyzing esterases. Anal Biochem 132:105–109.

Rosa E, Barata C, Damasio J, Bosch MP, Guerrero A (2006) Aquatic ecotoxicity of a pheromonal antagonist in *Daphnia magna* and *Desmodesmus subspicatus*. Aquat Toxicol 79:296–303.
Roy C, Grolleau G, Chamoulaud S, Riviere JL (2005) Plasma B-esterase activities in European raptors. J Wildl Dis 41:184–208.
Sanchez JC, Fossi MC, Focardi S (1997a) Serum "B" esterases as a nondestructive biomarker for monitoring the exposure of reptiles to organophosphorus insecticides. Ecotoxicol Environ Saf 38:45–52.
Sanchez JC, Fossi MC, Focardi S (1997b) Serum B esterases as a nondestructive biomarker in the lizard *Gallotia galloti* experimentally treated with parathion. Environ Toxicol Chem 16:1954–1961.
Sanchez-Hernandez JC (2006) Earthworm biomarkers in ecological risk assessment. Rev Environ Contam Toxicol 188:85–126.
Sanchez-Hernandez JC, Fossi MC, Leonzio C, Focardi S (1998) Use of biochemical biomarkers as a screening tool to focus the chemical monitoring of organic pollutants in the Biobio River basin (Chile). Chemosphere 37:699–710.
Sarkar A, Ray D, Shrivastava AN, Sarker S (2006) Molecular biomarkers: their significance and application in marine pollution monitoring. Ecotoxicology 15:333–340.
Satoh T, Hosokawa M (1995) Molecular aspects of carboxylesterase isoforms in comparison with other esterases. Toxicol Lett 82–83:439–445.
Satoh T, Hosokawa M (1998) The mammalian carboxylesterases: from molecules to functions. Annu Rev Pharmacol Toxicol 38:257–288.
Satoh T, Hosokawa M (2000) Organophosphates and their impact on the global environment. Neurotoxicology 21:223–227.
Satoh T, Hosokawa M (2006) Structure, function and regulation of carboxylesterases. Chem Biol Interact 162:195–211.
Satoh T, Taylor P, Borsron WF, Sanghani SP, Hosokawa M, La Du BN (2002) Current progress on esterases: from molecular structure to function. Drug Metab Dispos 30:488–493.
Schechter MS, Green N, LaForge FB (1949) Constituents of pyrethrum flowers. XXIII. Cinerolone and the synthesis of related cyclopentenolones. J Am Chem Soc 71:3165–3173.
Schulz R (2004) Field studies on exposure, effects, and risk mitigation of aquatic nonpoint-source insecticide pollution: a review. J Environ Qual 33:419–448.
Shan G, Stoutamire DW, Wengatz I, Gee SJ,. Hammock BD (1999) Development of an immunoassay for the pyrethroid insecticide esfenvalerate J Agric Food Chem 47:2145–2155.
Shan G, Leeman WR, Stoutamire DW, Gee SJ, Chang DPY, Hammock BD (2000) Enzyme-linked immunosorbent assay for the pyrethroid permethrin J Agric Food Chem 48:4032–4040.
Sharom MS, Solomon KR (1981) Adsorption and desorption of permethrin and other pesticides on glass and plastic materials used in bioassay procedures. Can J Fish Aquat Sci 38:199–204.
Sheets LP, Doherty JD, Law MW, Reiter LW, Crofton KM (1994) Age-dependent differences in the susceptibility of rats to deltamethrin. Toxicol Appl Pharmacol 126:186–190.
Shi D, Yang J, Yang D, LeCluyse EL, Black C, You L, Akhlaghi F, Yan B (2006) Anti-influenza prodrug oseltamivir is activated by carboxylesterase human carboxylesterase 1, and the activation is inhibited by antiplatelet agent clopidogrel. J Pharmacol Exp Ther 319:1477–1484.
Simpson SL, Micevska T, Adams MS, Stone A, Maher WA (2007) Establishing cause-effect relationships in hydrocarbon-contaminated sediments using a sublethal response of the benthic marine alga, *Entomoneis* cf. *punctulata*. Environ Toxicol Chem 26:163–170.
Singh BK, Walker A (2006) Microbial degradation of organophosphorus compounds. FEMS Microbiol Rev 30:428–471.
Soderlund DM, Casida JE (1977) Stereospecificity of pyrethroid metabolism in mammals. In: Elliot M (ed) Synthetic Pyrethroids. American Chemical Society, Washington, DC, pp 173–185.
Sogorb MA, Vilanova E (2002) Enzymes involved in the detoxification of organophosphorus, carbamate and pyrethroid insecticides through hydrolysis. Toxicol Let 128:215–228.

Sogorb MA, Vilanova E, Carrera V (2004) Future applications of phosphotriesterases in the prophylaxis and treatment of organophosporus insecticide and nerve agent poisonings. Toxicol Lett 151:219–233.
Solomon MG, Fitzgerald JD (1993) Orchard selection for resistance to a synthetic pyrethroid in organophosphate-resistant *Typhlodromus-Pyri* in the UK. Biocontrol Sci Technol 3:127–132.
Spurlock F, Bacey J, Starner K, Gill S (2005) A probabilistic screening model for evaluating pyrethroid surface water monitoring data. Environ Monit Assess 109:161–179.
Staudinger H, Ruzicka L (1924) Insektentotende stoffe. I-VI and VIII-X. Helv Chim Acta 7:177–458.
Stok J, Huang H, Jones PJ, Wheelock CE, Morisseau C, Hammock BD (2004a) Identification, expression and purification of a pyrethroid hydrolyzing carboxylesterase from mouse liver microsomes. J Biol Chem 279:29863–29869.
Stok JE, Goloshchapov A, Song C, Wheelock CE, Derbel MB, Morisseau C, Hammock BD (2004b) Investigation of the role of a second conserved serine in carboxylesterases via site-directed mutagenesis. Arch Biochem Biophys 430:247–255.
Straus DL, Chambers JE (1995) Inhibition of acetylcholinesterase and aliesterases of fingerling channel catfish by chlorpyrifos, parathion, and *S,S,S*,-tributyl phosphorotrithioate (DEF). Aquat Toxicol 33:311-324.
Sturm A, Wogram J, Segner H, Liess M (2000) Different sensitivity to organophosphates of acetylcholinesterase and butylcholinesterase from three-spined stickleback (*Gasterosteus aculeatus*) application in biomonitoring. Environ Toxicol Chem 19:1607–1615.
Sutherland TD, Weir KM, Lacey MJ, Horne I, Russell RJ, Oakeshott JG (2002) Enrichment of a microbial culture capable of degrading endosulphate, the toxic metabolite of endosulfan. J Appl Microbiol 92:541–548.
Sutherland TD, Horne I, Weir KM, Coppin CW, Williams MR, Selleck M, Russell RJ, Oakeshott JG (2004) Enzymatic bioremediation: from enzyme discovery to applications. Clin Exp Pharmacol Physiol 31:817–821.
Sweeney RE, Maxwell DM (1999) A theoretical model of the competition between hydrolase and carboxylesterase in protection against organophosphorus poisoning. Math Biosci 160:175–190.
Székács A, Bordás B, Hammock BD (1992) Transition state analog enzyme inhibitors: structure-activity relationships of trifluoromethyl ketones. In: Draber W, Fujita T (eds) Rational Approaches to Structure, Activity, and Ecotoxicology of Agrochemicals. CRC Press, Boca Raton, FL, pp 219–249.
Talcott RE (1979) Hepatic and extrahepatic malathion carboxylesterases. Assay and localization in the rat. Toxicol Appl Pharmacol 47:145–150.
Talcott RE, Denk H, Mallipudi NM (1979a) Malathion carboxylesterase activity in human liver and its inactivation by isomalathion. Toxicol Appl Pharmacol 49:373–376.
Talcott RE, Mallipudi NM, Fukuto TR (1979b) Malathion carboxylesterase titer and its relationship to malathion toxicity. Toxicol Appl Pharmacol 50:501–504.
Talcott RE, Mallipudi NM, Umetsu N, Fukuto TR (1979c) Inactivation of esterases by impurities isolated from technical malathion. Toxicol Appl Pharmacol 49:107–112.
Tang BK, Kalow W (1995) Variable activation of lovastatin by hydrolytic enzymes in human plasma and liver. J Clin Pharmacol 47:449–451.
TDC (2003) Insecticide Market Trends and Potential Water Quality Implications. San Francisco Estuary Project, San Mateo, CA.
Teh SJ, Deng D, Werner I, Teh F, Hung SS (2005) Sublethal toxicity of orchard stormwater runoff in Sacramento splittail (*Pogonichthys macrolepidotus*) larvae. Mar Environ Res 59:203–216.
Thompson HM (1993) Avian serum esterases: species and temporal variations and their possible consequences. Chem-Biol Interact 87:329–338.
Thompson HM (1999) Esterases as markers of exposure to organophosphates and carbamates. Ecotoxicology 8:369–384.
Thompson HM, Mackness MI, Walker CH, Hardy AR (1991a) Species differences in avian serum B esterases revealed by chromatofocusing and possible relationships of esterase activity to pesticide toxicity. Biochem Pharmacol 41:1235–1240.

Thompson HM, Walker CH, Hardy AR (1991b) Changes in activity of avian serum esterases following exposure to organophosphorus insecticides. Arch Environ Contam Toxicol 20:514–518.

Tippe A (1993) Are pyrethroids harmless? Evaluation of experimental data. Zentralbl Hyg Umweltmed 194:342–359.

Tyler CR, Beresford N, van der Woning M, Sumpter JP, Thorpe K (2000) Metabolism and environmental degradation of pyrethroid insecticides produce compounds with endocrine activities. Environ Toxicol Chem 19:801–809.

USEPA (1991) Methods for aquatic toxicity identification evaluations. Phase I Toxicity Characterization Procedures. EPA 600/6-91/003. Office of Research and Development, U.S. Environmental Protection Agency, Washington, DC.

USEPA (1992) Toxicity identification evaluation: characterization of chronically toxic effluents, phase I. EPA-600/6-91/005F. Office of Research and Development, USEPA, Duluth, MN.

USEPA (1993a) Methods for aquatic toxicity identification evaluations. Phase III Confirmation Procedures for Samples Exhibiting Acute and Chronic Toxicity. EPA 600/R-92/081. Office of Research and Development, USEPA, Washington, D.C.

USEPA (1993b) Methods for aquatic toxicity identification evaluations. Phase II Toxicity Identification Procedures for Samples Exhibiting Acute and Chronic Toxicity. EPA 600/R-92/080. Office of Research and Development, USEPA, Washington, DC.

USEPA (1996) Marine Toxicity Identification Evaluation (TIE): Phase I Guidance Document. EPA/600/R-95/054. Office of Research and Development, USEPA, Washington, DC.

USEPA (2002) Methods for measuring acute toxicity of effluents and receiving water to freshwater and marine organisms. EPA-821-R-02-021. Office of Research and Development, USEPA, Washington, DC.

USEPA (2004) Incidence and severity of sediment contamination in surface waters of the United States. National Sediment Quality Survey, 2nd Ed. EPA-823-R-04-007. Office of Science and Technology, Standards and Health Protection Division, USEPA, Washington, DC.

USEPA (2007) Sediment Toxicity Identification Evaluation (TIE) Phases I, II, and III Guidance Document. EPA 600/R-07/080. Office of Research and Development, Washington, DC.

Vijverberg HP, van den Bercken J (1990) Neurotoxicological effects and the mode of action of pyrethroid insecticides. Crit Rev Toxicol 21:105–126.

Vioque-Fernandez A, de Almeida EA, Ballesteros J, Garcia-Barrera T, Gomez-Ariza JL, Lopez-Barea J (2007a) Donana National Park survey using crayfish (*Procambarus clarkii*) as bioindicator: esterase inhibition and pollutant levels. Toxicol Lett 168:260–268.

Vioque-Fernandez A, de Almeida EA, Lopez-Barea J (2007b) Esterases as pesticide biomarkers in crayfish (*Procambarus clarkii*, Crustacea): tissue distribution, sensitivity to model compounds and recovery from inactivation. Comp Biochem Physiol C 145:404–412.

Wadkins RM, Morton CL, Weeks JK, Oliver L, Wierdl M, Danks MK, Potter PM (2001) Structural constraints affect the metabolism of 7-ethyl-10-[4-(1-piperidino)-1-piperidino]carbonyloxycamptothecin (CPT-11) by carboxylesterases. Mol Pharmacol 60:355–362.

Wadkins RM, Hyatt JL, Yoon KJ, Morton CL, Lee RE, Damodaran K, Beroza P, Danks MK, Potter PM (2004) Discovery of novel selective inhibitors of human intestinal carboxylesterase for the amelioration of irinotecan-induced diarrhea: synthesis, quantitative structure-activity relationship analysis, and biological activity. Mol Pharmacol 65:1336–1343.

Wadkins RM, Hyatt JL, Wei X, Yoon KJ, Wierdl M, Edwards CC, Morton CL, Obenauer JC, Damodaran K, Beroza P, Danks MK, Potter PM (2005) Identification and characterization of novel benzil (diphenylethane-1,2-dione) analogues as inhibitors of mammalian carboxylesterases. J Med Chem 48:2906–2915.

Wagner JM (1997) Food Quality Protection Act: its impact on the pesticide industry. Qual Assur 5:279–283.

Waller WT, Bailey HA, de Vlaming V, Ho KT, Hunt JW, Miller JL, Pillard DA, Rowland CD, Venables BJ (2005) Ambient water, interstitial water, and sediment. Society of Environmental Toxicology and Chemistry Press, Pensacola, FL, pp 93–114.

Ware GW, Whitacre DM (2004) The Pesticide Book, 6th Ed. Meister, Willoughby, OH.

Watanabe T, Shan G, Stoutamire DW, Gee SJ, Hammock BD (2001) Development of a class-specific immunoassay for the type I pyrethroid insecticides. Anal Chim Acta 444:119–129.

Wengatz I, Stoutamire DW, Gee SJ, Hammock BD (1998) Development of an enzyme-linked immunosorbent assay for the detection of the pyrethroid insecticide fenpropathrin. J Agric Food Chem 46:2211–2221.

Werner I, Deanovic LA, Connor V, de Vlaming V, Bailey HC, Hinton DE (2000) Insecticide-caused toxicity to *Ceriodaphnia dubia* (Cladocera) in the Sacramento-San Joaquin River Delta, California, USA. Environ Toxicol Chem 19:215–227.

Werner I, Deanovic LA, Hinton DE, Henderson JD, de Oliveira GH, Wilson BW, Krueger W, Wallender WW, Oliver MN, Zalom FG (2002) Toxicity of stormwater runoff after dormant spray application of diazinon and esfenvalerate (Asana) in a French prune orchard, Glenn county, California, USA. Bull Environ Contam Toxicol 68:29–36.

Werner I, Zalom FG, Oliver MN, Deanovic LA, Kimball TS, Henderson JD, Wilson BW, Krueger W, Wallender WW (2004) Toxicity of storm-water runoff after dormant spray application in a french prune orchard, Glenn County, California, USA: temporal patterns and the effect of ground covers. Environ Toxicol Chem 23:2719–2726.

West CW, Kosian PA, Mount DR, Makynen EA, Pasha MS, Sibley PK, Ankley GT (2001) Amendment of sediments with a carbonaceous resin reduces bioavailability of polycyclic aromatic hydrocarbons. Environ Toxicol Chem 20:1104–1111.

Weston D (2006) Temperature Dependence of Pyrethroid Toxicity: TIE Applications and Environmental Consequences. Society of Environmental Toxicological Chemistry, Montreal, Canada.

Weston DP, Amweg EL (2007) Whole sediment toxicity identification evaluation tools for pyrethroid insecticides: II. Esterase addition. Environ Toxicol Chem 26:2397–2404.

Weston DP, You J, Lydy MJ (2004) Distribution and toxicity of sediment-associated pesticides in the agriculture-dominated water bodies of California's Central Valley. Environ Sci Technol 38:2752–2759.

Weston DP, Holmes RW, You J, Lydy MJ (2005) Aquatic toxicity due to residential use of pyrethroid insecticides. Environ Sci Technol 39:9778–9784.

Wheelock CE, Severson TF, Hammock BD (2001) Synthesis of new carboxylesterase inhibitors and evaluation of potency and water solubility. Chem Res Toxicol 14:1563–1572.

Wheelock CE, Colvin ME, Uemura I, Olmstead MM, Nakagawa Y, Sanborn JR, Jones AD, Hammock BD (2002) Use of *ab initio* calculations to predict esterase inhibitor potency. J Med Chem 45:5576–5593.

Wheelock CE, Wheelock ÅM, Zhang R, Stok JE, Le Valley SE, Green CE, Hammock BD (2003) Evaluation of α-cyanoesters as fluorescent substrates for examining interindividual variation in general and pyrethroid-selective esterases in human liver microsomes. Anal Biochem 315:208–222.

Wheelock CE, Miller JL, Miller MG, Shan G, Gee SJ, Hammock BD (2004) Development of Toxicity Identification Evaluation (TIE) procedures for pyrethroid detection using esterase activity. Environ Toxicol Chem 23:2699–2708.

Wheelock CE, Eder KJ, Werner I, Huang H, Jones PD, Brammell BF, Elskus AA, Hammock BD (2005a) Individual variability in esterase activity and CYP1A levels in Chinook salmon (*Oncorhynchus tshawytscha*) exposed to esfenvalerate and chlorpyrifos. Aquat Toxicol 74:172–192.

Wheelock CE, Miller JL, Miller MJ, Phillips BM, Gee SJ, Tjeerdema RS, Hammock BD (2005b) Influence of container adsorption upon observed pyrethroid toxicity to *Ceriodaphnia dubia* and *Hyalella azteca*. Aquat Toxicol 74:47–52.

Wheelock CE, Shan G, Ottea JA (2005c) Overview of carboxylesterases and their role in metabolism of insecticides. J Pestic Sci 30:75–83.

Wheelock CE, Miller JL, Miller MJ, Phillips BM, Huntley SA, Gee SJ, Tjeerdema RS, Hammock BD (2006) Use of carboxylesterase activity to remove pyrethroid-associated toxicity to *Ceriodaphnia dubia* and *Hyalella azteca* in toxicity identification evaluations. Environ Toxicol Chem 25:973–984.

Wheelock CE, Nishi K, Ying A, Jones PD, Colvin ME, Olmstead MM, Hammock BD (in press) Influence of sulfur oxidation state and steric bulk upon trifluoromethyl ketone (TFK) binding kinetics to carboxylesterases and fatty acid amide hydrolase (FAAH). Bioorg Med Chem in press.

Williams FM (1985) Clinical significance of esterases in man. Clin Pharmacokinet 10:392–403.

Wilson BW, Henderson JD (1992) Blood esterase determinations as markers of exposure. Rev Environ Contam Toxicol 128:55–69.

Wogram J, Sturm A, Segner H, Liess M (2001) Effects of parathion on acetylcholinesterase, butyrylcholinesterase, and carboxylesterase in three-spined stickleback (*Gasterosteus aculeatus*) following short-term exposure. Environ Toxicol Chem 20:1528–1531.

Wortberg M., Jones G, Kreissig SB, Rocke DM, Gee SJ, Hammock BD (1996) An approach to the construction of an immunoarray for differentiating and quantitating cross reacting analytes. Anal Chim Acta 319:291–303.

Yamamoto R (1923) The insecticidal principle in *Chrysanthemum cinerariaefolium*. Part II and part III. On the constitution of pyrethronic acid. J Chem Soc Jpn 44:311–330.

Yamamoto R (1925) On the insecticidal principle of insect powder. Inst Phys Chem Res Tokyo 3:195.

Yang H, Carr PD, McLoughlin SY, Liu JW, Horne I, Qiu X, Jeffries CM, Russell RJ, Oakeshott JG, Ollis DL (2003) Evolution of an organophosphate-degrading enzyme: a comparison of natural and directed evolution. Protein Eng 16:135–145.

Yang WC, Gan JY, Hunter W, Spurlock F (2006a) Effect of suspended solids on bioavailability of pyrethroid insecticides. Environ Toxicol Chem 25:1585–1591.

Yang WC, Spurlock F, Liu WP, Gan JY (2006b) Effects of dissolved organic matter on permethrin bioavailability to *Daphnia* species. J Agric Food Chem 54:3967–3972.

Yang WC, Spurlock F, Liu WP, Gan JY (2006c) Inhibition of aquatic toxicity of pyrethroid insecticides by suspended sediment. Environ Toxicol Chem 25:1913–1919.

Yang ZP, Dettbarn WD (1998) Prevention of tolerance to the organophosphorus anticholinesterase paraoxon with carboxylesterase inhibitors. Biochem Pharmacol 55:1419–1426.

Zhang J, Burnell JC, Dumaual N, Bosron WF (1999) Binding and hydrolysis of meperidine by human liver carboxylesterase hCE-1. J Pharmacol Exp Ther 290:314–318.

Zhao GY, Rose RL, Hodgson E, Roe RM (1996) Biochemical mechanisms and diagnostic microassays for pyrethroid, carbamate, and organophosphate insecticide resistance/cross-resistance in the tobacco budworm, *Heliothis virescens*. Pestic Biochem Phys 56:183–195.

Index

B

C

M

O

P

MIX
Papier aus verantwortungsvollen Quellen
Paper from responsible sources
FSC® C105338

If you have any concerns about our products,
you can contact us on
ProductSafety@springernature.com

In case Publisher is established outside the EU,
the EU authorized representative is:
Springer Nature Customer Service Center GmbH
Europaplatz 3, 69115 Heidelberg, Germany

Printed by Libri Plureos GmbH
in Hamburg, Germany